JOURNEYS TO THE ENDS OF THE UNIVERSE

A guided tour of the beginnings and endings of planets, stars, galaxies and the universe

T0179310

JOURNEYS TO THE ENDS OF THE UNIVERSE

A guided tour of the beginnings and endings of planets, stars, galaxies and the universe

C R Kitchin
Hatfield Polytechnic Observatory

CRC Press
Taylor & Francis Group
Boca Raton London New York

CRC Press is an imprint of the
Taylor & Francis Group, an **informa** business

CRC Press
Taylor & Francis Group
6000 Broken Sound Parkway NW, Suite 300
Boca Raton, FL 33487-2742

First issued in paperback 2019

Typeset by BC Typesetting, Bristol BS15 5YS

No claim to original U.S. Government works

ISBN-13: 978-0-7503-0037-7 (hbk)
ISBN-13: 978-0-367-40323-2 (pbk)

British Library Cataloguing in Publication Data

Kitchin, Christopher R. (Christopher Robert), *1947–*
    Journeys to the ends of the universe: a guided tour of
    the beginnings and endings of planets, stars, galaxies and
    the universe.
    1. Universe
    I. Title
    523.1

US Library of Congress Cataloguing-in-Publication Data are available

**Visit the Taylor & Francis Web site at
http://www.taylorandfrancis.com**

**and the CRC Press Web site at
http://www.crcpress.com**

For Jennifer, in the hope, though not the expectation, that by the time she can read it, the remaining questions will have been answered.

Contents

	Preface	ix
Journey 1	An Inflated Idea of Ourselves	1
Journey 2	Strings to Tie a Universe Together	25
Journey 3	Phoenix Rising	44
Journey 4	The Nearest Star	72
Journey 5	A Study in Contrasts	88
Journey 6	Thar She Blows!	107
Journey 7	The Hottest Spots in the Universe	119
Journey 8	Cosmic Dustbins	147
Journey 9	To the Hub!	160
Journey 10	The Enigma Machines	172
Journey 11	She was the universe	184
	Bibliography	191
	Index	195

Preface

"Now, Axel," cried the Professor with enthusiasm, "now we are really going into the interior of the Earth. At this precise moment the journey commences."

So saying, my uncle took in one hand Ruhmkorff's apparatus, which was hanging from his neck; and with the other he formed an electric communication with the coil in the lantern, and a sufficiently bright light dispersed the darkness of the passage.

Journey to the Centre of the Earth
Jules Verne

In the spirit of Verne's Professor Liedenbrock, and in no lesser anticipation of the joys of discovery, this book undertakes forays into the hinterlands of modern science, exploring the limits of knowledge where scientific fact overtakes and merges with the wilder speculations of science fiction.

It provides a linked series of essays, suitable for the general reader as well as more specialist scientists, summarising the exciting topics currently developing apace within astronomy and physics and which are pressing outwards on the interfaces of science with belief.

The beginnings of galaxies, stars, planets, even life itself, are related back to the ravelled turmoil of the first few seconds and years in the life of the cosmos. The journey continues on past the ultimate fate of the solar system to probe the nature of the all-consuming fires of supernovae, taking in on the way the lesser deaths of more modest stars. The futures of galaxies, clusters of galaxies, superclusters of clusters of galaxies and so on then lead us towards the finale. It is there that some of the more bizarre musings of physicists and astronomers, suggesting possible destinies for the universe stretching billions of times its present age into the future, are encountered.

C R Kitchin
February 1990

Journey 1

An Inflated Idea of Ourselves

> *Unknown interlocutor: What was God doing before He created the Universe?*
> *St Augustine:*   *Inventing a Hell for people who ask such questions.*

"OK, but what was there before the beginning?" A child's not-so-naive question has long been able to bring the most erudite theologians and cosmologists to a shuddering halt in the midst of their most brilliant expositions on 'life, the universe and everything'. St Augustine's actual reply to the similarly motivated question cited above, though less quotable, was that "Time did not exist before the beginning of the universe".

This latter statement seems quite nonsensical at first sight, or even at second or third sight! It is an idea that is very difficult to express in ordinary language, since language itself arises from our everyday experiences and includes time as a universal and all-pervading property of the universe. Thus the word 'begin' itself assumes a continual flow of time, one specific point of which is identified as the start of something. A beginning to time is therefore undefined, since the concept of a beginning already requires time to exist. Nonetheless, St Augustine's real answer was very close indeed to some of the current ideas about the origin of the universe.

The journeys in this book will take the traveller along some weird and wonderful paths through modern scientific thinking, but on this first journey we shall encounter some of the most awesome, and perhaps shocking, ideas of all. We shall see that time, along with the three spatial dimensions, becomes just one of perhaps as many as 11 dimensions. These came into being out of nothing some ten to twenty thousand million years ago†, along with the fundamental forces of nature and the matter and radiation which make up our visible universe.

Before we encounter such ideas, however, we need to look at some of the features we observe in our universe to see how modern cosmologists are blowing away the mists which veil Creation, and to see how the role of God in the whole affair is being relegated at most to that of an uninvolved onlooker. This is necessarily only a brief and selective look at the subject of cosmology, and readers whose appetites are whetted by this

---

† Equations and mathematics generally are avoided in this book as far as possible. However, the numbers we encounter in astronomy are often extremely large, and sometimes extremely small. It is not very useful to write such numbers out in full: it is difficult to see readily that the number 'twenty-three thousand million million million million' differs by a factor of a million from the number 'twenty-three thousand million million million million million', without actually counting all the millions in each case. Equally, the difference remains difficult to spot with the numbers written out numerically: 23 000 000 000 000 000 000 000 000 000 and 23 000 000 000 000 000 000 000 000 000 000. Furthermore, it is just about impossible to remember such numbers if they need to be compared with other numbers at a later stage in the story. The commonly used convention of 'index notation' will therefore generally be employed for numbers over a million from now on. In this notation the number is written as a decimal with one figure before the decimal point which is then multiplied by however many thousands and millions may be required. The number of zeros in the thousands and millions is written as a superscript to the figure ten. Thus the above two numbers become

$$2.3 \times 10^{28} \quad \text{and} \quad 2.3 \times 10^{34}.$$

This is a much easier system to use, once you get used to it, but it is important to remember that a difference of only one unit in the index corresponds to a *factor* of ten difference in the numbers. Thus the numbers one hundred and one thousand would be expressed in this system as

$$10^2 \quad \text{and} \quad 10^3.$$

It is also possible to express very small numbers in this compact form. The index of ten is then negative. Numbers are again expressed as decimals with one figure before the decimal point and multiplied by a power of ten. The negative index though tells us that the number should in fact be *divided* by the thousands and millions corresponding to the index of ten. Thus the number 0.000 000 000 000 000 23 would be written as

$$2.3 \times 10^{-16}$$

and we can easily distinguish it from the number 0.000 000 000 000 000 0023, or

$$2.3 \times 10^{-18}$$

which is one hundred times smaller.

Figure 1.1   Star trails showing the darkness of the night sky between the stars.

account are referred to the bibliography for other books wherein to pursue their interests.

Observations made using huge and expensive telescopes which take many hours to grasp the few fleeting photons coming to us from the edges of the universe provide some clues to its origins. We shall examine some of these in due course.

One fundamental observation, however, does not need a telescope and can be made by anyone. It is that the sky is dark at night (figure 1.1). Johannes Kepler, whose main claim to fame is his decipherment of the laws of planetary motion, pointed out in 1610 that if stars were scattered at random in an infinitely old, infinitely large universe, then on looking out into space in any and all directions one would eventually see the surface of a star (figure 1.2). This would mean that the whole sky should have roughly the same surface brightness as the surface of an average star like the Sun. The Earth should then have a surface temperature of 5 000 to 6 000 K†. Since clearly the night sky (and the day-time sky for that matter) is nothing like so bright, nor the Earth's surface temperature

---

† Temperatures throughout this book are given in degrees Kelvin (absolute). These are the same as the Celsius (Centigrade) scale, but with the zero point shifted to −273 °C. For temperatures over a few thousand degrees, the two scales may be regarded as effectively the same.

Figure 1.2   The basis of Olbers' paradox.

so high, one or more of the original assumptions must be wrong.

Kepler's notion was rediscovered in the 19th century by Heinrich Olbers, and it is now generally and rather unfairly known as Olbers' paradox.

The explanation for the darkness of the sky lies in the universe not being infinitely old. Current estimates for its age lie between 10^{10} and 2×10^{10} years. When looking into the depths of space we also look backwards in time†. Eventually therefore we peer into regions where

---

†Light, radio waves, gamma rays and all other forms of electromagnetic radiation travel at a fixed speed of just under 300 000 kilometres per second (km s^{-1}) in a vacuum. This is often used to provide a unit for measuring distances of astronomical objects. The distance travelled by light in one year, known as the light year (ly), is

$$1 \text{ ly} = 9.5 \times 10^{12} \text{ km}.$$

Another common distance unit for astronomy is the parsec (pc) and is just over three times larger than the light year   (contd)

Figure 1.3   Helium in the universe.

the stars thin out because they are still forming, and so the requirement
for an infinite number of stars fails. Thus the simple observation of a dark
night sky tells us that the universe must have had an origin at some finite
time in the past.

There is another simple observation to be made by anyone (particularly if they have bought one of those aluminised helium-filled balloons at
a summer fete (figure 1.3)), though this fact is not as obvious as the
darkness of the sky: there is too much helium in the universe.

$$1 \, pc = 3.1 \times 10^{13} \, km = 3.3 \, ly.$$

This finite speed for light means that we see objects not as they are now, but as
they were when they emitted the light which we are now receiving. Thus we see
the Sun as it was eight minutes ago, for that is the time light takes to cover the
150 000 000 km between the Sun and the Earth. The next nearest star, Proxima
Centauri is 4.3 ly away, and we therefore see it as it was 4.3 years ago.
With more distant objects we observe them even further back in their past.
Undoubtedly, some of the stars which we still see in the sky have long since
ceased to exist. The most distant objects that we currently observe are galaxies
and QSOs (Journey 10) at distances of over 10^{10} ly. We therefore see them as they
were more than 10^{10} years ago, or over twice the present age of the Earth.

Currently, we find that the composition of most objects in the universe, such as stars, interstellar gas clouds, galaxies etc (but not planets like the Earth, which are special cases), is about 74% hydrogen and 24% helium. All the remaining one hundred or so elements amount to no more than 2% of the material in the universe.

We know that helium is produced inside most stars as they consume their hydrogen in nuclear reactions akin to those in a hydrogen bomb (see figures 4.1 and 4.2 in Journey 4). The amount of helium in the universe must therefore be increasing with time. However, all the visible stars in the universe, though they have been busily producing helium from hydrogen for some 10^{10} to 2×10^{10} years, would have converted only about 1% of that hydrogen into helium. Furthermore, most of this helium would still be locked deep inside the cores of those stars and hence not visible to us.

So in order to explain the observed amount of helium (24%), either the universe must have been created with it initially, or it was produced in some manner soon after the universe came into being. This latter possibility is now thought to be the true explanation. To follow this through, we must look in more detail at another fundamental observation underlying modern cosmology.

One of the results of the many years of observation of distant galaxies by astronomers has been the discovery that the further a particular galaxy is from us, the faster it is moving away (i.e. the universe is expanding). A galaxy three million light years distant from us would thus be moving away at between 50 and 100 km s^{-1} on average, while one that is six million light years away would normally be receding at between 100 and 200 km s^{-1}, and so on (figure 1.4). The relationship is known as the Hubble law, after its discovery in the 1920s by Edwin Hubble. The recession of the galaxies is also sometimes called the red-shift because of its effect on the radiation from the galaxies (shifting it to longer wavelengths).

This recession of the galaxies does not arise because we happen to occupy some especially noxious part of the universe from which everyone else is fleeing! In fact it is a general moving apart of all galaxies from each other. Observers in any other galaxy would also see a similar picture of galaxies receding from their own at velocities proportional to their distance. Going backwards in time, therefore, all the galaxies must have been closer together in the past, and some 10^{10} to 2×10^{10} years ago, they must all have been crammed 'on top' of each other in a very small volume indeed.

These and other observations led to the idea of what is now called the 'hot big bang' model for the origin of the universe. This, in essence, envisages all the matter and energy now observed in the universe originating from a very compact, extremely dense and hot, primaeval

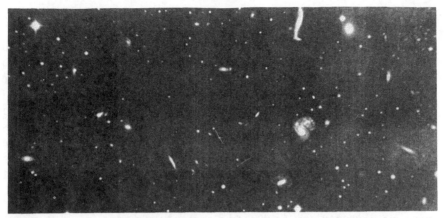

Figure 1.4 A cluster of galaxies in Hercules, about 500 million light years distant, and receding from us at about $10\ 000\ \text{km s}^{-1}$. (Reproduced by permission of the Palomar Observatory.)

fireball. The present recession of the galaxies, or expansion of the universe, is then just a relic of that original explosion.

The main stable particles produced in such an explosion would be protons and electrons: two of the three principal subatomic particles which form matter as we know it today. Protons and electrons would appear about one second after the start of the big bang, when the temperature had dropped to about 10^{10} K. Some tens of thousands of years later, as the temperature dropped to a few thousand K, these protons and electrons would then combine to form hydrogen atoms. The primordial matter would therefore seem likely to be pure hydrogen—a direct contradiction of our observation that about a quarter of it should be helium.

However, we have so far neglected to take into account that at temperatures in the region of 10^{10} K, the protons and electrons can react directly to produce neutrons, releasing on the way further particles called neutrinos. Neutrons are the third type of particle which along with protons and electrons form the chemical elements. Neutrinos are very unreactive particles, and are outside our everyday experience. They will, however, play fundamental roles in many of the processes which we will encounter on later journeys.

The key to understanding the production of helium lies in the fact that an opposing reaction can also take place. Neutrons and positrons (positive electrons) can combine to give back protons and neutrinos. The balance between the numbers of protons and neutrons under such conditions depends only upon the relative reaction rates for the two processes. These reaction rates can be determined in the laboratory and suggest that about one neutron would be in existence for every six protons.

About 100 seconds after the initial explosion, the temperature would have fallen to 10^9 K, and each neutron would combine with a proton to form a nucleus of deuterium ('heavy hydrogen'). Thus there would be about one deuterium nucleus to every five protons.

In due course the deuterium nuclei would combine in pairs to form helium nuclei, giving about one helium nucleus to ten protons. Thus when the free electrons eventually combined with these nuclei, the primordial matter would be formed of hydrogen and helium in the ratio of about ten hydrogen atoms to one helium atom. In terms of mass, that ratio is about 72% hydrogen to 28% helium.

This estimate is now rather higher than our observed ratio. However, we have neglected one further factor: that the neutron is an unstable particle with a half-life of about ten minutes. Some of the neutrons will therefore decay before they have time to form the deuterium nuclei (within which they are stable). Taking this last factor into account, precise calculations suggest that 23.6% of the primordial matter should be helium. Very good agreement with our observations at last.

The correct calculation of the amount of helium in the universe from the predictions of the hot big bang theory was a major advance in cosmology. This advance led to the elimination of the rival cosmological theory, the steady state theory, which had competed with the big bang theory in the 1950s and 1960s. The steady state theory envisaged the very large scale structure of the universe remaining constant. As the galaxies moved apart, new galaxies had to form in the voids to keep the overall density unchanged. This in turn required the continuous creation of matter to form those galaxies. The latter idea, of course, contravenes the principle of conservation of mass and energy†, but that is not the

---

† The principle of conservation of mass and energy is one of the fundamental laws of modern science. Briefly, it states

mass and energy can neither be created nor destroyed and in a closed system the total amount is constant.

In normal life, mass and energy are also separately conserved. Most so-called sources of energy are in fact just converting energy from one form to another. A fire, for example, converts the chemical energy stored in the coal or wood into radiant energy and the thermal energy of the surrounding fireplace, atmosphere etc.

However, as is now well known, it is possible to convert mass into energy, and this is the basis of nuclear power stations and atomic and hydrogen bombs, and, of course, the source of the Sun and other stars' energy. Thus mass is just another form of energy, and the relationship between the two (with c the velocity of light: 3×10^8 m s^{-1}) as deduced by Einstein is   (*contd*) drawback it might seem. Firstly, any theory of the universe is going to have to create matter somehow. Secondly, the required rate of the creation, one hydrogen atom per cubic metre every 10^{11} years, is so small that it is well within the accuracy to which the principle of the conservation of mass and energy has been proven.

The steady state theory, however, was unable to explain the hydrogen-to-helium ratio except by the *ad hoc* assumption that the newly created matter came into being with just the required ratio. Such an assumption in a theory renders it very unsatisfactory, essentially reducing the hypothesis to the proposition that 'God has decided it shall be so'. For a while some proponents of the steady state theory continued to try to uphold it, but it was soon dealt two further blows which were to prove fatal.

The first of these blows concerned radio galaxies, which are galaxies whose radio energy emissions are many times the total energy emission of a normal galaxy (see figure 1.5, and figure 2.4 of Journey 2). Because they are so powerful, they can be observed out to very great distances using radio telescopes. It is found that radio galaxies are much more common at large distances than closer to us. Since looking into the distance equates with looking into the past (see previous note), this implies that these galaxies were more frequent in the past than they are nowadays. Thus conditions in the past must have been different from those of today on a large scale. This is in direct contravention of the underlying assumption of the steady state theory, which requires the large scale structure of the universe to be unchanging.

The second observation was an even more conclusive blow to the steady state theory since not only was it inexplicable by the steady state theory, but it was predicted by the big bang theory. This was the observation in 1965 at Holmdel, New Jersey, by Arno Penzias and Robert

---

$$e = mc^2.$$

It is equally possible to convert energy into mass. The interchangeability of matter and energy is something that will be encountered many times on our journeys, and especially on this first one. It is useful therefore to have a unit to measure mass in terms of energy and the one most often used is the electron volt (eV). This is the energy gained by an electron in falling through an electrical potential of one volt. It is a very small amount of energy:

$$1 \text{ eV} = 1.6 \times 10^{-19} \text{ J}$$

(a single grain of sugar contains about 25 joules (J) of chemical energy). Nonetheless it is a useful unit when we are concerned with things on an atomic scale. The energy equivalents of some of the particles we have been talking about are:

electron or positron	500 000 eV	(0.5 MeV)
proton or neutron	1 000 MeV	

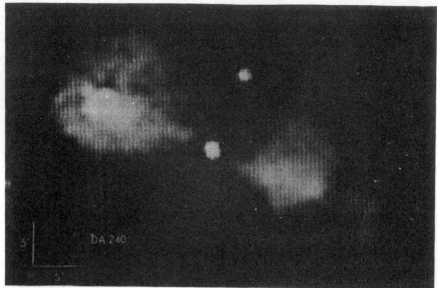

Figure 1.5   A Westerbork synthesised radio 'photograph' of the giant radio galaxy DA 240. (Reproduced by permission from *Nature* **250** 625 (1974), A G Willis, R G Strom and A G Wilson, © 1974 Macmillan Magazines Ltd. The Westerbork telescope is operated by the Foundation for Research in Astronomy with the financial support of the Netherlands Organisation for Scientific Research (NWO)).

Wilson of radiation in the microwave region (wavelengths of a few millimetres) coming uniformly from all parts of the sky.

George Gamow had predicted that just such low energy radiation should be filling the universe, nearly 20 years before its discovery. Unfortunately he predicted that the wavelength of the peak emission should be near a tenth of a millimetre, corresponding to the energy emitted by a body at a temperature of about 25 K ($-248\,°C$). Though the radiation level predicted by Gamow was very tiny by normal standards (the energy radiated from an ice cube would be 14 000 times greater!) and could not then be detected directly, it was still far too high to agree with other observations. For example, certain molecules in interstellar space were known to be at a temperature of only 3 or 4 K and that would be impossible if they were bathed by radiation at a temperature of 25 K.

The presence of background radiation is a fundamental requirement of any big bang model of the universe. It is just the remnant of the radiation produced by the original fireball, red-shifted from its unimaginable original levels by the expansion of the universe.

The apparent lack of such radiation was thus a severe drawback to the big bang theory in the 1950s and was one reason for the development of the competing steady state theory. Although in 1952 Walter Baade had shown that the galaxies were at least twice as far away as had been previously thought, and this change reduced Gamow's predicted radiation to a temperature of 5 to 10 K, there still appeared to be a conflict with observation which the big bang theory had difficulty explaining.

This was all changed by Penzias and Wilson's observations. They were actually trying at the time to detect signals from two of the early communications satellites: Echo I and Echo II. These 'spacecraft' were just huge aluminised balloons in low Earth orbits which reflected any radio signal directed at them. Unlike modern communications satellites, they did not receive, amplify and re-broadcast the signal in a narrow beam; the reflected signal, by the time it reached the surface of the Earth again, was thus very weak indeed, and Penzias and Wilson had to try very hard to reduce all interfering noise sources to the absolute minimum. Try as they might, however, they could not get rid of one problem—a faint hiss of noise at a radio wavelength of 70 mm which came uniformly from all parts of the sky. The intensity of this radiation corresponded to that coming from a body at a temperature of 3.5 K ($-269.7\,°C$).

Almost simultaneously and only a few miles away at Princeton, Robert Dicke was independently rediscovering Gamow's predicted background radiation on the basis of his own version of a hot big bang model for the origin of the universe.

The theoretician and the observers soon combined their ideas to suggest that in this faint hiss of radio noise we were listening for the first time to the 'birth pangs' of the universe. Since then many observations have confirmed these ideas and refined the measurements. The radiation is now known to correspond very closely to that which would be emitted by a body at a temperature of 2.7 K ($-270.5\,°C$), and it is generally known as the microwave background radiation (figure 1.6).

The discovery of the microwave background radiation was a triumph for the hot big bang theoreticians, and for a while it seemed as though only minor details remained to be explained before we should completely understand how our universe came into being.

More recently, however, a number of problems have become apparent, and attempts to solve them have led to some radical ideas about what may have occurred in the very early stages of the big bang. The latest, but almost certainly not the last, such development is termed chaotic inflation. It does provide explanations for many puzzling aspects of the universe. After a small fraction of a second though, the various modern ideas all lead to much the same model for the expansion and subsequent evolution of the universe as the original hot big bang theory.

The first problem was obvious from an early stage, but theoreticians managed to avoid it by starting their models not from the instant of

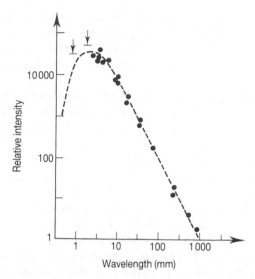

Figure 1.6   The microwave background radiation (dots: observations, broken line: theoretical curve for 2.7 K).

creation, but from a small fraction of a second later. The problem is often
termed the singularity problem. It is that a simple reversal of the present
expansion of the universe would lead back to an object of infinite density
and zero volume at the birth of the universe.

Such a phenomenon is termed a singularity by physicists and is not
capable of description by our present scientific laws. The explanation for
the existence of such a singularity can thus only be in terms of an 'act of
God', which is no explanation at all. The singularity problem is the most
fundamental of the problems of modern cosmology since it also involves
the question of 'where did all the matter and radiation come from?' It still
has not been completely resolved, but there is at least an indication of
where the answer may lie, and this is in the concept of quantum
fluctuations.

The two fundamental ideas underlying modern physics are general
relativity and quantum mechanics. Many attempts have been made to
join these two ideas into a single quantum theory of gravity, but so
far without success. Until that is done, the solution to the singularity problem will remain somewhat speculative. Nonetheless, quantum
mechanics alone can suggest a possible solution to the singularity
problem.

Reference has already been made to the idea that in a closed system the
total amount of matter and energy is fixed: the principle of the conservation of mass and energy. In the vernacular this could be expressed as
'you get nowt for nowt' and it is certainly confirmed by our everyday

experiences. However, the idea relies upon our ability to assess exactly how much matter and energy there is in such a closed system. If our measurements of these quantities contain uncertainties, then their total could vary without our being aware of it.

Now a basic result known as Heisenberg's uncertainty principle states that the act of measurement itself will disturb that which we are trying to measure. There will thus inevitably be uncertainties in measurements, no matter how careful and precise we may try to be.

As an example of such a disturbance, imagine trying to measure the position and velocity of a very light particle such as a single proton. There would be various ways of doing this: passing it through a magnetic field, catching it in a photographic emulsion, illuminating it with light etc. However, the magnetic fields, silver bromide grains, photons etc, whilst being affected by the proton (for it is thus that we hope to measure the proton's position and speed), will also disturb the proton in their turn. The more precisely we manage to measure one of the quantities, the worse will be the disturbance to the other.

There is thus a theoretical limit to the precision with which we can measure two properties of an object simultaneously. This limit is very tiny, which is why the effect is not noticeable in the macroscopic world. If the position of the proton were known to within one micron (a millionth of a metre), the greatest precision to within which its velocity could simultaneously be found would be about 0.4 m s^{-1}. If its position were known to a nanometre (1 nm $= 10^{-9}$ m), the velocity could only be found to within 400 m s^{-1} and so on.

If the temperature of an object were to be reduced to absolute zero, then it might be expected that the component particles would become motionless. However, if that were the case then we would only need to determine the particle's positions in order to know both their energy (zero) and position to any required degree of accuracy. Such a double measurement would violate the uncertainty principle. So some residual and unremovable energy remains with the particles even at zero temperature. This energy is called the zero-point energy. It serves to keep the energies and positions of the particles sufficiently uncertain for the requirements of the uncertainty principle. In case this seems a rather abstruse and esoteric consideration, it does have observable macroscopic effects. For example, the zero-point energy of helium is sufficient to keep it liquid and to make it impossible to solidify even at zero temperature.

Position and velocity (or momentum) are one pair of quantities whose mutual precision of measurement is limited by the uncertainty principle, but not the only ones. In particular, mass/energy and time are another such pair.

To return to our previous consideration of the principle of conservation of mass and energy, we must now include the effect of the uncertainty principle. We can thus only say that in a closed system the total amount of mass and energy is constant to within the limits of our measurements. The more precisely we know the instant of the measurement, the less precisely will we know the total amount of mass and energy.

If we were to measure the total amount of mass and energy to an uncertainty of 1.7×10^{-27} kg (the mass of a proton), then the uncertainty in the timing of that measurement will be at least 4×10^{-24} seconds. To put this another way, in our closed system, a proton could come into being from nowhere. Providing that the proton then disappeared again in less than 4×10^{-24} seconds, we would be quite unable to detect it by any current technique *or any technique which might be invented in the future*. The brief presence of this proton would not therefore upset the principle of the conservation of mass and energy.

In fact, a single proton could not appear in this manner because other conservation laws would be broken. But a proton and an anti-proton could come into being simultaneously without any such constraint since the positive charge of the proton would be balanced by the negative charge of the anti-proton and so on. Providing then that the two particles annihilated each other in less than 2×10^{-24} seconds (half the time required for the single proton, because the energy involved has doubled), the principle of conservation of mass and energy would again not be violated.

A proton and an anti-proton annihilating each other would normally result in the emission of a pair of very high energy gamma rays. In the situation just envisaged, however, nothing would result. The energy released when the two particles combine goes to 'pay off' the 'energy debt' caused by their formation. The net effect is for a proton and an anti-proton to flicker into being for a fleeting fraction of a second and then to disappear again leaving no direct evidence of their existence. In a similar way any particle and its anti-particle—electron and positron, neutron and anti-neutron, two photons (a photon is its own anti-particle) etc—could come into being. The only constraint on the process is the requirement that the particles disappear again in a time smaller than that imposed by the uncertainty principle for the continuing conservation of mass and energy.

Such pairs of particles are termed virtual particles (or quantum fluctuations) by reason of their non-detectability and are thought to fill the universe, 'empty' space as well as locales such as the Earth which contain many real particles.

Since virtual particles are not detectable, you may well be wondering what all the fuss is about and why we do not just ignore them. The answer is that though they are not *directly* detectable, their presence does produce effects which can be observed. Virtual particles provide the mechanism for the transmission of the forces of nature, and their presence near black holes results in black holes not being truly black but radiating energy continuously (Journey 8). Similar fluctuations in electromagnetic and gravitational fields result in small but measurable changes in the energies of electrons in atoms (the Lamb shift) and a force between two separated metal plates (the Casimir effect).

We may seem to have strayed a long way from the initial question of the origin of the universe and the solution to the singularity problem, but in fact we now have sufficient background information to enable us to make further progress.

We have seen that a proton and anti-proton pair can flash into existence for up to 2×10^{-24} seconds before they must disappear again. An electron and positron being about 2 000 times less massive could remain in existence 2 000 times longer. A pair of visible light photons, with an energy 200 000 times smaller still, could last for nearly 10^{-15} seconds. A pair of radio wave photons with wavelengths of 300 000 km could last for up to one second. Edward Tryon continued the logic of this argument in the early 1970s to point out that a quantum fluctuation with zero energy could remain in existence forever.

The significance of this last point is that within the universe, the gravitational potential energy is *negative†*, whilst that of the particles and radiation is *positive*. It is thus possible for the net energy of the whole universe (the sum of the total gravitational energy and the total mass energy) to be zero. Thus in turn it is possible for the entire universe to have originated as a zero-energy quantum fluctuation. As such it would come into being with a finite size and so avoid the problem of the singularity.

As we shall see shortly, in response to the solution to another problem, the initial quantum fluctuation would not contain all the present contents of the observed universe. Indeed, everything that we now see was probably contained within an initial volume no more than 10^{-35} m across and with less than the equivalent of 10^{-8} kg of matter in it. This small volume was then inflated to at least the size of the currently

---

† It is easy to see that potential energy must be negative if we imagine all the matter in the universe spread out uniformly, with infinite distances between adjacent particles. Then no particle would experience a gravitational force from any other particle. The gravitational potential energy of the universe would thus be zero. If the particles are then imagined to fall towards each other, gravitational energy would be released. The gravitational potential energy of the universe would hence reduce, from zero towards negative values. Our actual universe, within which matter is clearly not spread out to infinite distances, must therefore have a negative potential energy.

observable universe (10^{26} m), creating the observed contents of the universe on the way.

One consequence of such an origin for the universe is that it must contain a finite amount of matter and energy, and that the gravitational field may be sufficient eventually to halt the observed expansion and to cause it to collapse again. The universe could then disappear back into the nothingness that was prior to the original quantum fluctuation. Confirmation, though not proof, of this idea would come from showing that the universe would one day stop expanding.

There is an observation that may be made to determine whether or not the gravitational field of the universe is strong enough to halt its expansion. That observation is of the mean density of the universe (see also Journey 11). The matter currently observed in planets, stars, nebulae, galaxies, etc, if spread throughout the universe, would amount to about one atom per ten cubic metres. This density is too small by about a factor of one hundred to halt the expansion. Yet even so it may be counted as a very close miss indeed to the required (or critical) density. Other than with the type of origin just outlined there is no *a priori* reason for the actual density of the universe to be within many orders of magnitude of the critical density: 10^{-53} or 10^{2071} etc times the critical density would be just as acceptable.

To make the coincidence even more striking, there is evidence that there may be many times the amount of matter that we can see directly in the universe which is as yet unobserved. One such line of evidence is the determination of the masses required for clusters of galaxies to be stable, which suggests that this 'missing mass' may amount to 10 to 50 times the amount of visible matter. Should this latter observation be confirmed, then the mean density of the universe is probably the same as the critical density to within the uncertainties of the measurements. Even before such confirmation, many cosmologists believe that the universe will prove to be exactly the critical density simply because of the improbability of the observed density being so close to the critical density without some effect constraining them to equal each other.

Clues to the way in which the transition from the quantum fluctuation to the present universe occurred come from two other cosmological problems called the flatness and the isotropy problems, respectively.

The flatness problem concerns the geometry of space. It is possible for space to be closed (i.e. in the form of a four-dimensional sphere), to be flat (Euclidean), or to be open (figure 1.7). A pair of parallel lines extended far enough would then converge, remain at a constant separation or diverge, respectively. Since there are an infinite number of possible closed geometries, and an infinite number of open geometries, but only one Euclidean geometry, it is overwhelmingly probable that space should be curved. Nonetheless, all the observational evidence suggests that the

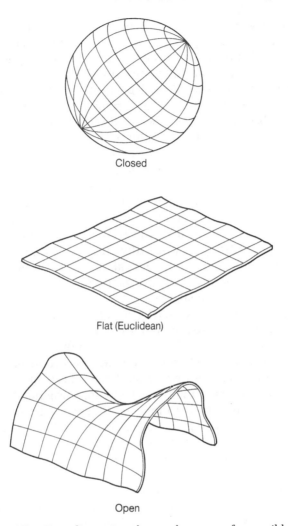

Closed

Flat (Euclidean)

Open

Figure 1.7 Two-dimensional analogues of possible four-dimensional geometries for the universe.

geometry of the universe is indeed very close to being flat. The flatness problem is then to find an explanation for why the universe fits this very special case.

The isotropy problem is simply stated. It is that the universe on large scales is highly isotropic (i.e. it has the same properties in all directions). Thus we find for example that the microwave background radiation has almost the same temperature wherever we look. The small variation that is found occurs because of the Earth's motion. The radiation appears very slightly warmer in the direction towards which we are headed (near the constellation of Virgo), and very slightly cooler on the other side of the sky (near Pisces). Similarly, when absorption by dust clouds etc within our own galaxy is taken into account, there are roughly equal numbers of galaxies to be found in all directions, and so on.

To see why isotropy should be a problem, let us concentrate on the background radiation. This is the remains of radiation from the primaeval fireball at a time about 300 000 years after the initial explosion. That radiation has by now been red-shifted by the expansion of the universe to a thousand times its original wavelengths.

Now the properties of the matter and radiation at a particular place and time can only be influenced by events sufficiently close for inter-actions to have occurred since the beginning of the universe. Since the velocity of light provides an upper limit to the rate at which such influences can spread outwards, the background radiation originating from a given point 300 000 years after the big bang can only possibly be affected by events occurring within about 300 000 light years of that point. Such a limit is often called the horizon for that point, and the distance to it, the horizon distance.

Considering the microwave background radiation coming to us from opposite sides of the sky, we find that when it originated, those two points would have been separated by nearly a hundred times their horizon distances. Thus there could have been no interaction between the two regions, and their past histories would have been quite unconnected. There is therefore no reason why the radiations originating from the two regions should bear any resemblance to each other at all. Yet the emissions from the two regions are almost identical, and this lack of variation constitutes the isotropy problem.

The postulated explanation for both the flatness and isotropy problems is via a very brief period of very rapid expansion very shortly after the universe came into being. This period of expansion, often termed the inflationary period, probably occurred when the universe was between 10^{-33} and 10^{-30} seconds 'old' and increased the size of the universe by at least a factor of 10^{40}, and possibly by as much as $10^{1\,000\,000}$. This last is such a huge number that to write it out in full would need a book four times longer than this one and containing just one followed by a million zeros (figure 1.8).

At the instant when the inflation started, the size of the region which was causally connected would have been about 10^{-35} metres. Inflation boosted its size dramatically, at least up to the order of a metre in size, and possibly very much larger. The speed of this expansion was much faster than that of light, but since it was space itself that was expanding and not an object moving through space, the speed limit of the speed of light imposed by special relativity was not contravened.

After this moment of hectic activity, the inflation came to an end, and

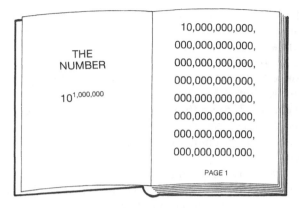

Figure 1.8   The number 10 to the power 1 000 000.

the 'normal' expansion due to the big bang took over. Over the next 1.5×10^{10} years or so, this expansion pushed the size of that little bit of the original quantum fluctuation that was causally connected (10^{-35} m) up to at least the present observed size of the universe: 10^{26} m.   The two problems of flatness and isotropy now have a simple explanation.

Isotropy occurs because the whole of the observed universe *did* originate from a small, uniform region that was causally connected at about 10^{-33} seconds. The faster-than-light inflationary expansion of *space* then led to the apparent lack of connection between regions at the time of the origin of the microwave background radiation because signals can only travel *through* space at the speed of light.

The flatness is equally easily explained now. The huge expansion during the inflationary period would cause whatever geometry the universe started out with to look very close to flat over the 'small' scale of the size of the visible universe. As an illustration imagine the Earth expanded in size by a factor of a million (a far smaller expansion factor than that experienced by the universe as a whole). Now we know that the Earth is approximately spherical, and classic observations such as watching a ship come up over the horizon can prove this. However, the observations are not easy, as evidence by the long persistence of the idea of a flat Earth. They would be far more difficult to make on such a very much larger body, and for almost all practical purposes we could treat the surface as though it were flat.

The idea of this early period of ultra-rapid inflation also solves other problems that have not so far been mentioned. One of these concerns monopoles, subatomic particles that are very massive (about 10^{16} times the mass of a proton, or about the mass of this full stop: .). They contain a single magnetic pole, north or south, and are the equivalent in magnetic terms of the much more familiar electrical point charges such as electrons. Monopoles are predicted to be produced in vast quantities in the early stages of the big bang and should still be around today. Yet despite many attempts, no one has yet managed to find one. The problem is thus to explain where the monopoles have gone.

The answer to the monopole problem suggested by current cosmological ideas is that the initial inflation means that the universe has expanded very much more than was thought even a few years ago. The monopoles are thus spread out over a very much larger region than used to be expected, and our chances of finding one in the visible universe are small indeed.

Inflation also provides a solution to yet another problem, the edge problem, but our discussion of that must be postponed for a while.

We have thus seen that a period of very rapid expansion very shortly after the start of the big bang can solve some otherwise intractable problems. But we have not touched on why such a period of expansion might occur. Let us go back to our original quantum fluctuation. We have concentrated on that small fraction of it which became the universe we now inhabit. But at 10^{-33} seconds there would be many such regions in which very different conditions might hold because they are outside each other's horizons. André Linde who first suggested the idea, likens this stage to a chaotic space 'foam'.

The composition of the foam at this stage is of many types of high energy field. A region containing a field with a very high potential energy will have most of its energy locked up in that field, and little in the form of the kinetic energy of 'ordinary' particles.

Now the rate of acceleration of the expansion of the universe, according to general relativity, depends upon its energy density. In the region under consideration the energy is mostly the potential energy of the field, and the energy density is thus negative (see previous note on gravitational potential energy) and the acceleration therefore positive.

The potential energy of the field decreases only very slowly as the region expands, and so the expansion occurs with very rapidly increasing speed. The period of inflation lasts until the field has dropped down to its minimum value and the net energy density reduced to zero or to positive values. Thus the inflation will continue longest in those regions which start out with fields of highest potential energy (i.e. furthest away from their minima), and such regions will expand to the largest sizes.

When the field reaches its lowest value, it oscillates about that minimum. The energy of the field is converted into high-energy subatomic particles. The previously largely empty region thus becomes filled with radiation and particles.

In the case of the region which became 'our' universe, these particles and radiation would amount to at least all the matter and energy that we detect at present in the visible universe. The continuing much slower and decelerating 'normal' expansion of the universe, which we still observe today, would then lead to the production of the microwave background radiation, the conversion of hydrogen to helium, and so on, to form the universe as we now know it.

But our universe has come from only one of the regions. What of the others? In some cases the fields may roll down to the same minimum and there will be a similar universe which may be larger or smaller depending on the original potential energy. In others, however, the fields may end up at different minima. Then different laws of physics will apply and the universes will be very different from our own.

To see why this is so we need to look again at the fields filling the early universe. The four forces which exist in the universe: the familiar gravitational and electromagnetic forces and the probably rather less familiar weak and strong nuclear forces, each arise from the effects of a field. Thus in a different region, with the fields at different minima, the forces of nature could be different both in number and in kind.

The differences between regions, however, could be even more fundamental. We are familiar with the three dimensions of space, and have become used to the idea that time in some way is a fourth dimension. However, in attempting to unify ideas about fundamental particles and forces, physicists have found that the most successful theories require ten or eleven dimensions. If they are right, why do we only perceive four dimensions?

The idea behind the answer may be illustrated by a simple garden hosepipe. This is a three-dimensional object, but if we look at it from some distance (several hundred metres) then it will appear as just a curved line: a one-dimensional object—its other two dimensions have become too small to be seen (see figure 1.9). In fact, though, the other dimensions are still there. Each point on the line is actually a small circle. In the physicist's terms, the extra dimensions have been compactified.

This is what is thought to have happened to the other six or seven dimensions in our universe—they have become compactified to a scale (10^{-35} to 10^{-32} m) which is far too small to be directly detectable by any of our present techniques. Each point in our four-dimensional space-time is thus not a true geometrical point, but is a tightly curled knot of the other dimensions†.

Now it is only in a universe with three spatial dimensions that planets

---

† This idea is often known as superstring theory, from the proposal that the cause of the compactification is a myriad of almost infinitesimal elastic strings which fill the universe. These collapse down the compactified dimensions until they are too small for us to detect. They also produce the particles within the universe by spinning, vibrating and interacting with each other.

Figure 1.9 A hosepipe as a one-dimensional and a three-dimensional object.

and electrons have stable orbits, enabling matter as we know it to exist and ultimately allowing life and ourselves to evolve. There is no reason, however, why the dimensions should follow the same pattern of compactification in other regions, and they may well have anything from one to eleven macroscopic dimensions. Clearly, such regions would be very different from our own universe.

What happens at the interface between two regions which differ in some or all of the ways discussed above? The properties and laws applying in one region must somehow change into those applying in the other region. If such an interface, or edge, occurred in the visible part of our universe, then it would show up as a region of extreme anisotropy.

As we have seen, however, the visible universe is highly isotropic. The lack of observation of such an edge to the universe is the edge problem to which reference was made earlier. Its explanation on the inflationary scenario is simple: the period of rapid inflation increased the size of the universe to such an extent that the visible portion is only a very small fraction of the whole, and the chance of our visible universe containing an interface to another region is therefore vanishingly small.

Our current picture of the universe is thus a strange one. The visible universe is just a small portion of a very much larger region, the whole of which originated via inflation from a small fragment of a quantum fluctuation (figure 1.10). Removed from us by a fifth, sixth, seventh,

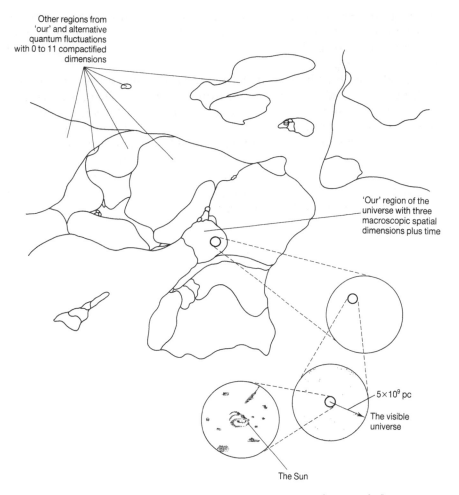

Figure 1.10   A high schematic visualisation in pseudo-3D of the current multiple-'space-foam' many-dimensionally-chaotic ideas on the nature of the universe.

eighth etc dimension, there are other regions of varying sizes, made up of different numbers of macroscopic dimensions and containing greater or fewer or different forces. Some of these regions might adjoin each other, others might be forever separated.

The answer to why our universe has three dimensions is straightforward: if it had any other number of spatial dimensions, then we would not have evolved to be able to see it. This argument is known as the anthropic principle and is not a true explanation. The anthropic principle, however, does show that in a large collection of regions with varying numbers of compactified dimensions, it is only in those with six or seven compactified dimensions and which are large enough to remain in existence for several billions of years that observers of our type would be likely to emerge.

A final thought. All this activity has so far originated from the chaotic distribution of fields in a single quantum fluctuation. There is no reason for that to have been the only such fluctuation. In fact, there are likely to have been many, maybe even an infinite number of similar events. It is even conceivable that a sufficiently advanced civilisation could create the necessary conditions, 10^{96} kg m^{-3} at a temperature of 10^{24} K over a region at least 10^{-30} m, to start their own family of universes off on their hectic expansionary journey. Perhaps with this last idea we have finally arrived at a role for God in our universe.

Journey 2

Strings to Tie a Universe Together

Se non e vero, e ben trovato!

Italian Proverb   (Even if it is not true, it is well invented!)

In the first journey, we found that to explain observations such as the high level of isotropy of the microwave background radiation, an enormous expansion (inflation) of the universe must have occurred very early on in its evolution. Another consequence of this inflation was that the geometry of space-time would be flattened until on the 'small' scale of the visible universe it differed little from Euclidean geometry. Other 'lumps and bumps' in the early universe would similarly have been smoothed away by the inflation.

The part of the universe that we can now see would thus have been in an extremely smooth and homogeneous state from a very early stage. Yet now we see a very different picture. Matter in the universe is distributed in clumps—small, very dense concentrations (stars and planets etc) interspersed through large volumes of low density material (the inter-stellar medium). On larger scales the inhomogeneities continue. The stars cluster into galaxies, which in turn form clusters and then clusters of clusters, all separated by large voids.

How could such a 'lumpy' universe have developed from the initial smooth state? Our next journey takes us off on a search for a solution to this problem.

Let us start by looking in more detail at the galaxies themselves. It comes as a surprise to modern minds, accustomed to the idea that the Milky Way galaxy is just one of millions of similar star systems, to find that it was only in the first decade of this century that galaxies were actually recognised as being composed of stars (figure 2.1), and not of thin hot gas like the nebula in Orion. Indeed, it was not until 1923 that they were unequivocally proven to be at distances of millions of light years and thus to lie outside our own galaxy.

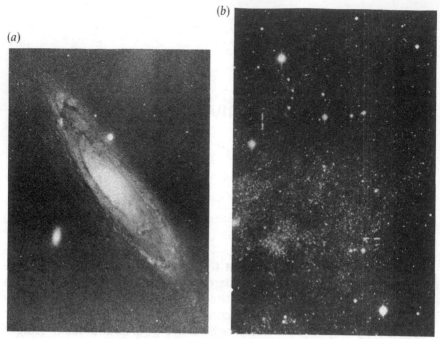

Figure 2.1   (a) The Andromeda galaxy (M31) and (b) parts of its spiral arms resolved into stars. (Reproduced by permission of the Palomar Observatory.)

These denizens of the space outside our own galaxy are now known to be many and various. There are galaxies like our own, beautiful spirals composed of 10^{11} stars or more and permeated by gas and dust clouds, and elliptical galaxies, looking like the nuclei of the spiral galaxies but sometimes as much as ten times larger (figure 2.2). These two types are sometimes called normal galaxies, but as we shall see later (Journey 9) they may still be very odd by our everyday Earth-centred standards.

Then there are such strange objects as the quasars (Journey 10). Quasars were first found in the 1960s as intense radio sources. When observed in the visible they appeared as blue star-like objects (hence their name, derived from **qua**si **s**tellar **r**adio **s**ource). Only a few per cent of the quasars now known have the predominating radio emission of those found at first. They are among the brightest objects in the universe, up to 10 000 times brighter than a normal galaxy, and visible out to distances of 10^{10} ly or more. Yet all their energy emanates from a region perhaps no larger than our own solar system. The most popular current explanation for them envisages their energy being supplied as material cascades down into a black hole containing perhaps 10^9 times the mass of our Sun.

(a)

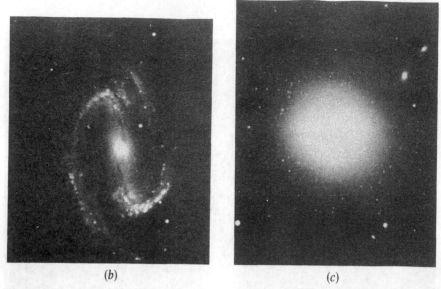

(b)   (c)

Figure 2.2   (a) M81 (NGC 3031). (Lick Observatory photograph.) (b) NGC 1300. (Reproduced by permission of the Palomar Observatory.) (c) M87 (NGC 4486). (Reproduced by permission of the National Optical Astronomical Observatories.)

Intermediate between quasars and spiral galaxies are the Seyfert galaxies (figure 2.3). They have small, intensely bright centres to their nuclei but otherwise are fairly normal looking spiral galaxies.

Then there are the radio galaxies (figures 1.5 and 2.4). These may be emitting many times more energy at radio wavelengths than in the visible. Usually the optical counterparts appear as fairly normal giant elliptical galaxies, but sometimes they seem to be undergoing violent contortions and explosions and/or emitting high velocity jets of material. Frequently the radio emission comes not from the optically visible galaxy, but from regions on opposite sides of it and separated from it by up to ten million light years.

At least as strange are the BL Lac objects with their variable polarized emission and the narrow emission line galaxies (NELGS) whose name explains their main distinguishing property. There are LINERs, and star-burst galaxies, mega-masers and Markarian galaxies; the list could be extended almost indefinitely.

As always, however, the greatest attention is paid to the few peculiar or spectacular objects that stand out from the crowd. In fact all the galaxies and other objects mentioned so far are fewer in total than the most common type of galaxy: the small elliptical or irregular galaxy

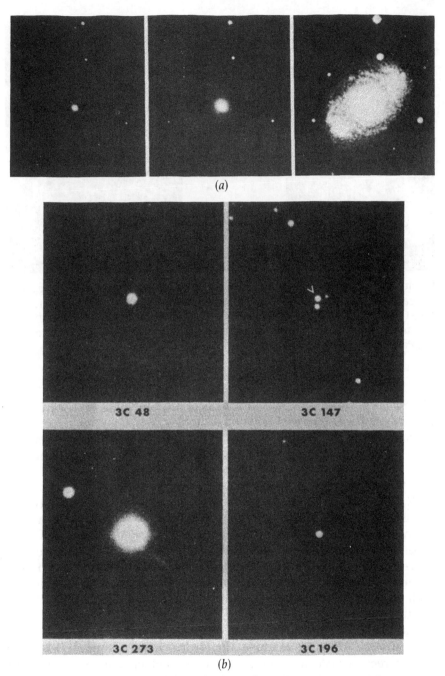

Figure 2.3   (a) NGC 4151, a Seyfert galaxy. Exposures increasing to the right. (b) Quasars. (All photographs reproduced by permission of the Palomar Observatory.)

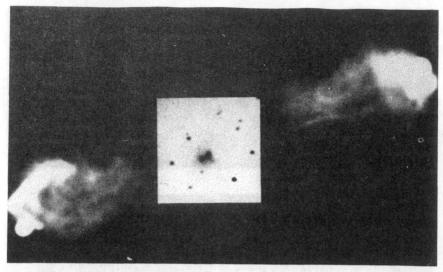

Figure 2.4   Cyg A radio source: a radio pseudophotograph with a negative optical image of the central galaxy superimposed. (Radio image courtesy of NRAO/AUI, observers P A G Scheuer, R A Laing and P A Perley; optical image courtesy of L A Thompson, University of Illinois; montage courtesy of J M Pasachoff, Hopkins Observatory.)

(figure 2.5). These unremarkable objects are too small and faint to be seen beyond a few tens of millions of light years. They may, however, contain more material than the rest of the galaxies put together, perhaps even sufficient to form the 'missing mass' required to close the universe (Journeys 1 and 11).

It is easy enough with even quite small telescopes to observe the individual galaxies (figure 2.6). The larger scale structure of the universe, however, is not so directly obvious. When we look into space we see objects at all distances projected onto the celestial sphere and thus superimposed one on top of another. To study the structure of the universe, therefore, it is necessary to find a method of determining the distances of the objects so that we can separate out those that are close to each other in space from those that simply lie near to the same line of sight.

With galaxies, some indication of their distance comes from their angular sizes. Thus, for example, the Hydra cluster (figure 2.7) is fairly readily recognisable as a group of associated galaxies from a direct photograph. Most clusters of galaxies, however, are not so obvious. Even with the Hydra cluster there are images which might be either small galaxies belonging to the cluster or large galaxies much further away.

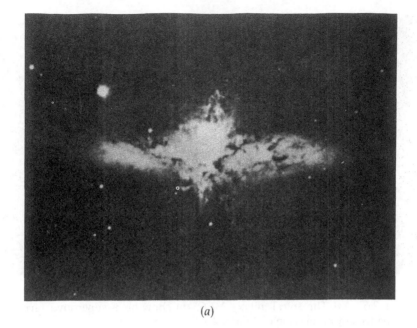

(a)

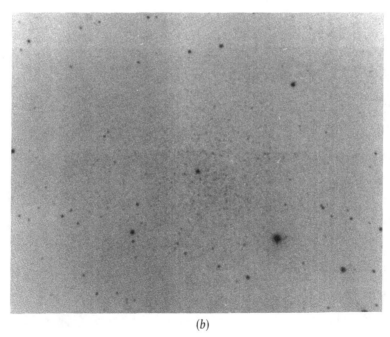

(b)

Figure 2.5   Dwarf irregular and elliptical galaxies. (a) M82 (NGC 3034) and (b) Leo II (negative photograph). (Reproduced by permission of the Hale Observatories.)

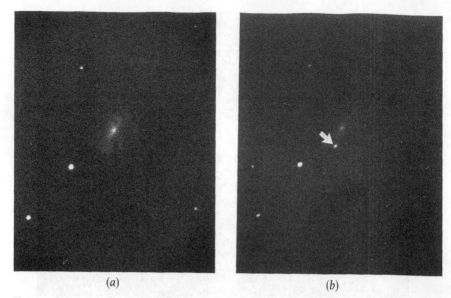

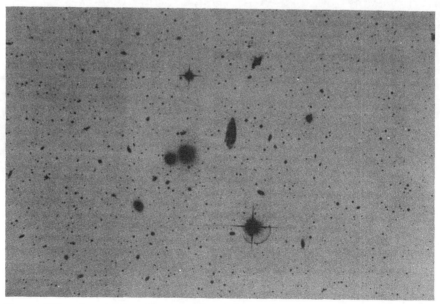

Figure 2.6   (a) The spiral galaxy M66, (b) showing a supernova (arrowed). (Reproduced by permission of R Forrest.)

Figure 2.7   A negative photograph of the centre of the Hydra I cluster of galaxies. (Reproduced by permission of the European Southern Observatory.)

Luckily we can get an estimate of the distance of a galaxy from its spectrum. As we saw on the last journey, the origin of the universe left the material in it exploding outwards at high velocities, and this remains to this day as the general recession of the galaxies. The Hubble law derived from the expansion gives an average recessional velocity of a galaxy as $15-30$ km s^{-1} for every million light years of distance away from us†. Hence observing the spectrum of the galaxy enables its red-shift to be measured. The red-shift in turn gives its recessional velocity, which may finally be converted into distance via the Hubble relationship.

In the 1950s, the whole of the northern sky was photographed using the newly built 1.2 m Schmidt telescope at Mount Palomar. The plates from that survey were then examined by George Abell to find clusters of

---

† As mentioned in Journey 1, this effect is often also called the red-shift of the galaxies. The latter name derives from the fact that when an object moves away from us, the features in its spectrum are moved to longer wavelengths. Thus in the visible spectrum the lines are moved towards the red.

This is a special case of the Doppler shift. There is a similar effect for objects moving towards us, when features are moved to shorter wavelengths—a blue-shift. The fractional change in wavelength is given by the velocity of the object divided by the velocity of light (300 000 km s^{-1}), and is thus normally quite small. Sodium D lines (the well known yellow sodium light) have rest wave-lengths of 589.0 and 589.6 nm (1 nm $= 10^{-9}$ m). In a source moving radially at 100 km s^{-1} (i.e. more than ten times faster than a spacecraft re-entering the Earth's atmosphere), these would change to 589.2 and 589.8 nm if it were moving away and to 588.8 and 589.4 nm if it were moving towards us. The Doppler shift only applies to velocities or components of velocities along the line of sight. There is no effect on wavelength due to motion across the line of sight (except for velocities close to the speed of light when relativistic effects do introduce a transverse Doppler shift).

The features in a spectrum arise from the atoms and molecules producing the light. They are usually called spectrum lines because light enters most spectro-graphs via a slit, resulting in linear bright or dark regions wherever there is a greater or lesser emission of radiation than normal (figure 2.8). Each atom, ion (an atom that has lost one or more electrons, or, more rarely, has gained an extra electron) or molecule will absorb or emit radiation only at certain specific wavelengths. The resulting pattern of emission or absorption lines is character-istic of that particular atom, ion or molecule, and of no other. The patterns can therefore be used to identify the presence of the atom etc in the source; thus the two bright yellow lines of sodium already mentioned. Hydrogen by contrast exhibits a series of lines (known as the Balmer series) starting in the red region (the Hα) at 656.3 nm, continuing to shorter wavelengths with Hβ at 486.1 nm, Hγ at 434.0 nm, Hδ at 410.2 nm and so on. Such spectrum lines are only produced by a thin gas. High pressure gases, liquids and solids emit or absorb across very wide wavelength ranges, and their spectra are therefore generally called continuous.

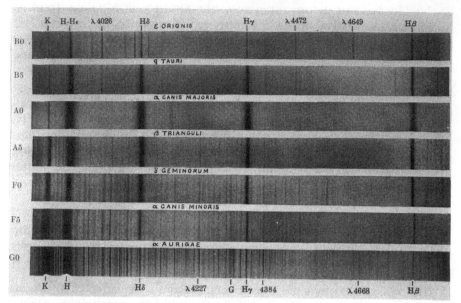

Figure 2.8   Typical stellar spectra, showing absorption lines due to hydrogen and other elements. (Reproduced by permission of the Royal Astronomical Society.)

galaxies. He found over 2 700 such clusters with at least 50 bright galaxies per cluster. In some cases, such as the Virgo group, there were many thousands of galaxies in the cluster. In fact, the majority of galaxies seem to belong to one cluster or another, though the local group containing the Milky Way, the Andromeda galaxy, and 15 other galaxies is too small to be called a cluster by Abell's definition.

The natural question to ask next, having found galaxies to occur within groups, is whether or not there are even larger structures. Are the clusters of galaxies linked in any way, or do they occur at random throughout the universe? The answer to that question has only become clear in the last decade: yes, there are superclusters.

This last result, obtained by Jack Burns and others, has required a vast amount of time and effort using the largest optical telescopes. The reason why it has taken so long is that though quite small instruments can photograph the galaxies directly, often with many images on a single plate, it takes many hours on the largest telescopes to obtain the spectrum of a single galaxy. Only recently with the development of the very efficient charge-coupled device (CCD) detectors have sufficient spectra become available to give a reasonable representation of the distribution in space of galaxies and clusters within 1.5×10^9 ly of the Earth.

Using the positions of some 650 of the brighters clusters of galaxies, Burns and his group chose a cluster at random and then looked for another cluster within 1.2×10^8 ly. If one or more such clusters were found, then the search was extended to look for another cluster within a similar range of that cluster and so on. In this way over 100 super-clusters were found, with less than one chance in a million that the distribution was due to chance and not to a real physical relationship.

The shapes of the superclusters tend to be filamentary: long lines of associated clusters strung out through space, often with small clusters and individual galaxies bridging the gaps (figure 2.9). Between the superclusters are voids, regions of space up to 3×10^8 ly across, almost devoid of matter apart perhaps from a few poorish clusters and individual galaxies.

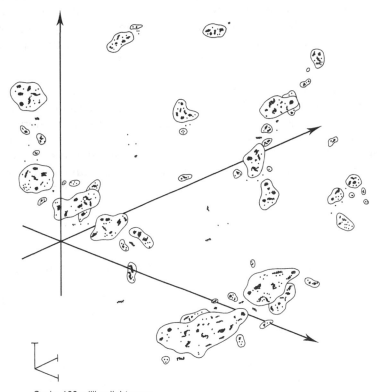

Scale: 100 million light years

Figure 2.9   Schematic view of very large scale structure in the universe, showing clusters of galaxies, the linear form of the superclusters (clusters of clusters of galaxies), and the voids between. (Individual galaxies are shown about $10 \times$ their true scale size.)

Does the hierarchy continue? Are there super-superclusters? As yet no answer can be given to those questions. The existence of numerous superclusters has only become apparent recently, and then only by using the best of today's instruments and techniques. Perhaps the appropriately named Hubble space telescope will provide the answer, since it is projected to be able to detect objects a hundred times fainter (and therefore up to ten times further away) than we can achieve with ground-based instrumentation. However, at the time of writing the launch of that telescope is six months away and dependent upon the continuing success of NASA's problem-ridden space shuttle programme.

Let us return then to examining how this lumpy structure for the universe could have arisen from the extremely smooth conditions after the inflationary period. Early ideas suggested that fluctuations would arise and behave either isothermally or adiabatically or perhaps both.

In an isothermal fluctuation, the *temperature* remains constant within the fluctuation, and equal to that of the rest of the universe. Isothermal models suggest that the galaxies would form first, only later joining together to form clusters. The superclusters, however, then present a problem because there has been insufficient time since the origin of the universe for their formation.

In an adiabatic fluctuation, the *energy* within the fluctuation remains constant and so the temperature increases as the fluctuation condenses. Adiabatic fluctuations lead to the formation of the superclusters first, to be followed by fragmentation into clusters and eventually into galaxies. Unfortunately, the large scale temperature fluctuations inherent in adiabatic models would cause the microwave background radiation to be equally lumpy—something we know from Journey 1 not to be the case.

A way out of this latter problem has been suggested: that the bulk of the matter in the universe is not in the form of visible matter, but in the form of neutrinos or some other weakly interacting particle. This is attractive at first sight for, again from Journey 1, we know that there is suspected to be a large amount of 'missing mass' in the universe. Neutrinos could provide that mass. Unfortunately again, however, the predictions based upon this theory do not agree with the observations. In particular, the theory predicts that the galaxies would not form until less than 10^{10} years ago, and we know that there are stars older than that in globular clusters and galactic nuclei.

So how could the galaxies, clusters, superclusters etc have come into being? While a long way from being established as the correct answer, current ideas are concentrating on the possible role played by cosmic strings.

In Journey 1 we encountered superstring theory: each point in our 'normal' four-dimensional space-time universe may actually be formed from a tiny knot of another six or seven dimensions on a scale of 10^{-33} m or thereabouts. These compactified dimensions may arise through the action of uncountable, almost imperceptible, elastic strings permeating the universe, whose interactions may also produce the observable subatomic particles. Now, cosmic strings have somewhat similar dimensions (10^{-32} m) in some directions, but in other directions they may stretch across the whole universe. They may also have as much mass as an entire cluster of galaxies. Thus, apart from the congruence of names, cosmic strings and the strings of superstring theory are totally different entities.

To understand the idea of cosmic strings we need to return to the very early stages of the big bang, to 10^{-35} seconds into the life of the universe, before the start of the inflationary expansion. As we have seen, the temperature of the universe was then incredibly high; but more importantly, the fields which later fuel the inflation and whose symmetry breaking as they roll down to their minima results in the forces such as gravity and electromagnetism, were still at very high energies. Those fields would subsequently decline in energy. In regions of the universe separated by more than the horizon distance, however, there would be no reason for the fields to decline in the same way. Within a single such region, therefore, the fields would be uniform, but at interfaces they must somehow change from one orientation to another. In so doing, the fields across the interface will retain a portion of their original energy.

We can illustrate this process more clearly by an analogy. Imagine a surface like 'bubble-plastic' sheeting: numerous hillocks set out regularly on a plain (figure 2.10). Further let us imagine that there are many small marbles dashing around on this surface, each attracted to its neighbours by an elastic force. When the marbles have high energies they will be distributed at random, as often on the tops of the hillocks as in the

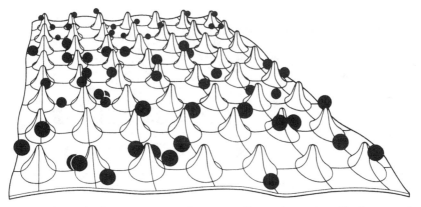

Figure 2.10   A high energy state: the mutually attracting 'marbles' are moving rapidly and with many different individual energies.

valleys. As their energies reduce, however, they will tend to settle in the valleys. The attractive forces between them will then ensure that they come to rest in a regular grid pattern with each marble in the same relative position within its valley as its neighbours (figure 2.11). The same process will happen in other, distant, parts of the bubble-plastic universe, but the positions taken up by the marbles within the valleys is likely to be quite different from those in the region first considered (figure 2.12). At the interface between two such regions therefore, the attractive forces between the marbles will displace them from the valleys, sometimes to the tops of the hillocks. Thus the marbles along the interface will retain a portion of their initial high energy.

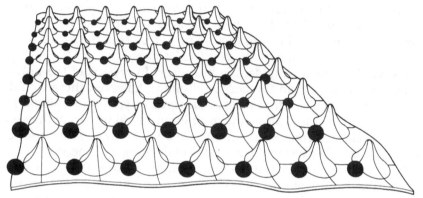

Figure 2.11   One of the infinite number of minimum energy positions for the mutually attracting 'marbles'.

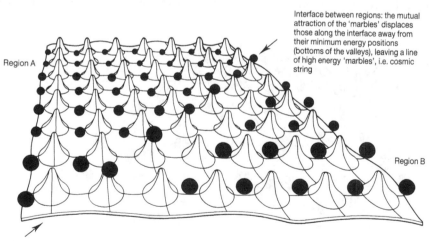

Region A

Interface between regions: the mutual attraction of the 'marbles' displaces those along the interface away from their minimum energy positions (bottoms of the valleys), leaving a line of high energy 'marbles', i.e. cosmic string

Region B

Figure 2.12   A line of cosmic string at the interface between two regions with differing minima.

The high energy marbles along the interface in our analogy would be the cosmic strings of the bubble-plastic universe. In the real universe then, the cosmic strings are regions of trapped energetic fields from the very early stages of the universe: long thin tubes of high energy or 'false' vacuum.

Within a string some of the properties of the early universe would be retained. In the highest energy strings, which would also be the thinnest, we would still find that the strong and weak nuclear forces and electromagnetic forces were indistinguishable from each other, just as in the early stages of the universe. The strings would be extremely massive; typically each metre of string could have a mass of 10^{21} kg. Just 6 km of string would contain as much mass as the whole Earth. They would also be extremely taut, 10^{37} newtons (N) or so, a force sufficient to stretch a steel cable 8 000 000 000 km in diameter to twice its normal length!

Only infinitely long strings or closed loops of string can therefore exist. The ends of a finite length of string would quickly snap together and it would disappear. The maximum size possible for a loop of string would be the relevant horizon distance for the instant of its formation. Since the horizon distance increases with time, loops would have tended to be smaller on average in the past.

Two strings crossing each other can break and rejoin at their intersection to form new structures, and tension will then cause waves to propagate along the strings at close to the speed of light. We can thus picture the universe as crossed by a network of wriggling lengths of cosmic string continually crossing and interacting, and with closed loops forming and budding off to move independently through space.

The closed loops of cosmic string should increase in both number and size as the universe ages. Perhaps fortunately, loops can disappear as well as form. The wave motion of the string generates gravitational waves† and in the closed loops these remove the energy of the string until it

---

† Gravity waves are emitted whenever an object changes its position or the distribution of material within itself. The effect is easy enough to observe in some ways. The Moon, for example, in travelling around the Earth causes the tides to follow it. What is less obvious is that the effect of the changing position of the Moon or other object is not instantaneous, but propagates outwards at the speed of light. Thus if the Sun were suddenly to vanish completely, the Earth would continue to follow its normal orbit for about eight minutes: the time it would take for the change in the gravitational field to travel from the Sun to the Earth. Any change in position or structure of an object, whether periodic or not, thus results in the consequent changes in the gravitational field, the gravity waves, being emitted and carrying energy with them. The exact nature and properties of the gravity waves must be predicted by gravitational theory for they have yet to be detected directly, and these predictions vary between differing theories.

shrinks and vanishes. Typically about 10 000 oscillations of the string are required for it to disappear. Given that the oscillations are moving at near light speeds, the time for a single oscillation to occur in *years* is comparable with the size of the loop in *light years*. Thus only loops a million light years or so across could have survived to the present day from soon after the origin of the universe.

To return to the original question of how galaxies etc could have formed in the smooth wake of the inflationary period of the early universe, we can now add to our picture of the conditions at the time (about 10 000 years from the origin of the universe) this all-pervading network of massive cosmic strings. Loops of string some 100 ly across with masses in the region of 10^{39} kg (10^9 times the mass of the Sun, or about 1% of the mass of a large galaxy) would act as nuclei to concentrate nearby material and thus cause the formation of galaxies. Larger loops would have sufficient mass to gather both the smaller loops and their protogalaxies into the clusters. Yet larger loops would have resulted in superclusters.

All these loops would have decayed because of their emission of gravitational radiation long ago. Computer simulations of the formation of loops from oscillating strings, however, suggest that before disappearing they would leave the galaxies and clusters etc partitioned into exactly the hierarchical filamentary distribution that we saw earlier probably to be the true pattern.

Cosmic strings, should they exist, can provide us with a mechanism for disrupting the extremely smooth conditions occurring after the inflation of the universe. Thus might the galaxies have formed, ultimately of course leading to the formation of stars, planets and ourselves.

It is a natural question to ask whether such strange entities as the superstrings might not have other effects, and whether perhaps we could detect them directly today. One prediction of the theory is that the hierarchy of galaxies–clusters–superclusters should stop with the superclusters. The answer to the early question of whether super-superclusters exist would then be *no*. This is because the loops required to form the superclusters would be about 10 000 ly across. That size is comparable with the horizon distance at the time of the superclusters' formation. The larger loops required for the formation of larger structures, therefore, could not have existed then. However, as we have seen, observations are as yet insufficient to decide whether this larger scale structure exists or not.

Another possibility is that loops of a million light years or more across, and those strings which span the universe, should still be in existence today. Such loops and strings are predicted to induce abrupt changes in the intensity of the microwave background radiation by about one part in 100 000. Detection of such small variations is beyond the capabilities of present-day equipment, but improvements in detectors and the COBE (Cosmic Background Explorer Spacecraft) should allow us to look for them within a few years. Similarly the gravitational waves produced by the decaying loops are too faint to be picked up by the gravity wave detectors available at present, but may be observable soon.

Another possible candidate for a loop of cosmic string could be the so-called 'great attractor'. Our own galaxy and local cluster, along with other clusters and superclusters within a range of about 150 Mly (1 Mly = 10^6 ly) are all moving at about 700 km s^{-1} towards a point 150 to 300 Mly away in the direction of Hydra. A mass 10^{16} times that of the Sun would suffice to accelerate these clusters to such a velocity over the lifetime of the universe—but this huge mass has yet to be found. A giant supercluster of galaxies might be one answer, but equally it could be the result of a moving loop of cosmic string a few tens of millions of light years across.

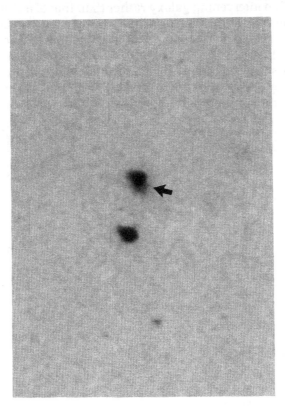

Figure 2.13   A gravitational lens in Ursa Major: the double quasar 0957 + 561 A and B. The galaxy which is acting as the lens is arrowed. (Reproduced by permission of A Stockton, University of Hawaii.)

What we *can* look for today is the gravitational lens effect of a string on the radiation from galaxies etc lying behind it. Light is affected by gravity fields in that its path is bent from a straight line (Journeys 8 and 10). The effect is small: a beam of light just skimming the surface of the Sun would be deflected by just 1.75 seconds of arc (0.0005 degrees). The strings are so massive though, that they should produce observable effects even if they are many millions of light years away.

The gravitational lens is different from an optical lens in that the deflection of the light beam reduces rather than increases at greater distances away from the optical axis (figure 10.2). It does not produce a genuine focused image, therefore, but rather a blurred and split image. We should thus look for multiple images of a single galaxy (or quasar etc) if we wish to detect cosmic strings.

Several such multiple images of both quasars and galaxies have been found (figure 2.13). Most, however, seem to be due to the gravitational lens effect of an intervening galaxy rather than that of a string.

In just one case, at the time of writing, is there a possibility of evidence of a string. This is for a pair of quasars found in Leo and known unromantically as $1146 + 111$ B and C. They are almost certainly two images of a single quasar as their red-shifts and spectra are identical. However, they are separated by over 150 seconds of arc in the sky, a far greater separation than for any other known pair of gravitational

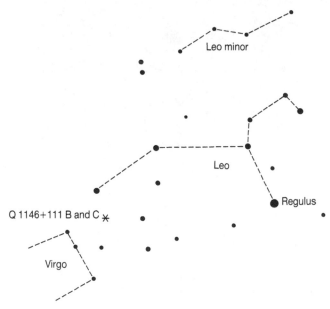

Figure 2.14   Leo and the possible cosmic string.

images. It is just possible that a cluster of galaxies between us and the quasar would have a sufficiently large gravitational field to produce this degree of separation. But such a cluster should also be near enough to be visible itself, and yet has not so far been found. Similarly a massive black hole could produce the same effect but should also be detectable. Since a cosmic string should extend across the sky, support for its existence would come if an alignment of several such widely separated double images could be found, and an active search for them is currently taking place, but so far with only negative results. Thus at the moment we are left with only the tantalising possibility that our first piece of cosmic string may be found along a line of sight just below the Lion's tail (figure 2.14).

Journey 3

Phoenix Rising

Glendower:
At my nativity,
the front of heaven was full of fiery shapes,
of burning cressets, and at my birth
the frame and huge foundation of the Earth
shaked like a coward.

Henry IV part 1
W Shakespeare

On our first two journeys we have had to travel far in space and time and delve into some obscure crannies of scientific thinking in order to reach our destinations. Now, for a brief while, we can return to where matter behaves in more familiar ways, and go in search of how and where the Earth, the Sun and other stars, and other planets came into being.

We know, for example from the ratios of some long-lived radioactive elements to their decay products, that the Earth is about 4.55 aeons (thousand million years) old. It is generally assumed that the Sun and the rest of the solar system are of a similar age.

The universe as a whole, as we saw on Journey 1, is probably about three times older still. So somehow, about 10 aeons after the big bang, the material now forming the solar system must have been brought together into a small region and shaped into the Sun and planets.

Now we do not have just the origin of the Sun about which to speculate, but also that of all the stars, because the Sun is a typical star. Extending our journey in this way might seem to add to the problem. However, stars are still being born today, so we can actually observe many of the details of the process directly. Even with this observational assistance though, it has not been straightforward to deduce how stars form. Only recently, with the advent of high quality infrared observations, have some of the details been established and many puzzles still remain. Current ideas have developed tortuously and in fits and starts.

We can, however, ignore the sequence in which the ideas were formed, and just take the logical path of following through the birth processes of a star from beginning to end.

'Space', the volume which contains the planets, Sun and stars, and through which they move, is usually regarded as a vacuum, but this is very far from being correct. Close to the Earth, where most spacecraft operate, there are in fact some 10^{14} atoms, ions and electrons per cubic metre (for comparison there are about 3×10^{25} per cubic metre of the atmosphere at sea level). Spacecraft in orbits below about 400 km are thus moving through gas so dense that it causes an appreciable drag. Their orbits therefore eventually decay and they re-enter the atmosphere and (usually) burn up.

We can sometimes see this gas, and the dust particles associated with it, as the zodiacal light (figure 3.1). This is a faint glow to be seen on moonless clear nights stretching up from the point on the horizon where the Sun has set or from where it will rise, and following the line of the ecliptic†.

Moving away from the Earth, the density of the interplanetary gas falls, and if we leave the solar system entirely, then we find that the interstellar medium is even thinner: about 10^6 atoms, ions and electrons per cubic metre.

Most of the galaxy therefore is not empty space, but is filled by a very thin gas which has roughly the same composition as the Sun and other stars. That is to say, the interstellar medium is mostly hydrogen and helium with a small admixture of the other elements. Here and there, however, we find thickenings of the interstellar medium, sometimes by factors of between a thousand and a million. These thickenings we call interstellar nebulae, and they may be up to a hundred light years across (figure 3.2).

Now although the gas in these nebulae is still much less dense than the space between the planets in the solar system, even a nebula 'only' a light year in diameter will contain as much material as the whole solar system, including the Sun. The larger nebulae therefore contain thousands or millions of solar masses of material.

---

† The ecliptic is the path of the Sun around the sky throughout a year. The 12 constellations of the zodiac are distributed along it and the Moon, planets and asteroids are usually to be found within them. The constellation containing the Sun at the time of a birth gives rise to the concept of 'birth signs' and the whole inanity of astrology. Through the movement of the Earth's axis in space (precession), the conventional astrological birth signs no longer correspond to the true position of the Sun. Thus on March 21st when the Sun is supposed to enter the constellation of Aries according to astrology, it is actually in the middle of Pisces and so has another 30 days travelling to do before reaching Aries.

Figure 3.1   The zodiacal light. (Reproduced by permission of Octopus Group Ltd, original source unknown.)

It is in such giant nebulae, often called giant molecular clouds (GMCs) for reasons that will become apparent shortly, that stars and planetary systems are forming today.

There are, of course, many types of interstellar nebula. They may be divided into emission, reflection and absorption nebulae according to whether we detect them from their own emissions, from the light they scatter from a nearby star or from their absorption of the light of the more distant stars behind them (figure 3.3).

Alternatively, nebulae may be divided according to their modes of origin. Planetary nebulae and supernova remnants we shall encounter later (Journeys 5 and 7). The nebulae of interest now are the ones called HI and HII (aitch-one and aitch-two) regions. These names come from the predominating state of ionisation of hydrogen within the nebulae. A common way of symbolising an ion is to use the chemical symbol followed by a roman numeral to indicate the number of electrons that the atom has lost. Since the neutral atom is given a roman 'I', the required numeral is actually one more than the number of lost electrons. Thus on this system, neutral hydrogen is HI, and ionised hydrogen (a bare proton since hydrogen has only one electron) is HII. An HI region is thus composed mostly of neutral hydrogen, and most of this will have combined into hydrogen molecules. An HII region is composed mostly of ionised hydrogen: protons and electrons moving independently.

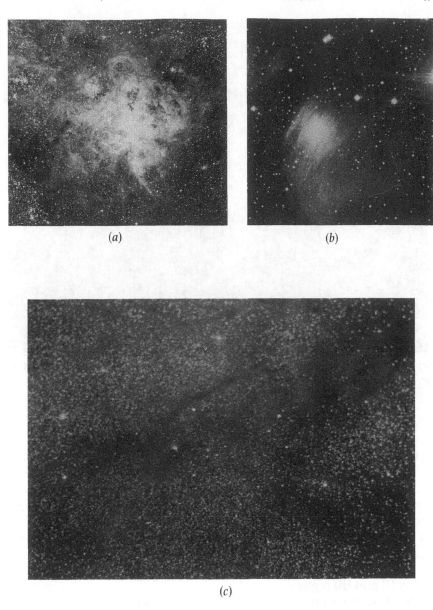

Figure 3.2   Varieties of interstellar nebulae. (*a*) An HII region. The Tarantula nebula in the Large Magellanic Cloud. (Reproduced by permission of the Royal Observatory, Edinburgh.) (*b*) Reflected nebula around Merope in the Pleiades. (Reproduced by permission of the Royal Astronomical Society.) (*c*) An absorption nebula: the Taurus molecular cloud. (Reproduced by permission of the Carnegie Institute of Washington.)

Figure 3.3   The Horsehead nebula (an absorption nebula). (Reproduced by permission of the Royal Observatory, Edinburgh.)

The HI regions are cold (10 to 100 K), while the HII regions are hot (10 000 K or so). Additionally, the HI regions have about 1% of their mass as small (100 to 1000 nm) dust particles, which have condensed out of the 4% or so of the nebula's material that is in the form of elements heavier than hydrogen and helium. The dust particles are very much more efficient at absorbing radiation than the more abundant gases and so the denser parts of the nebula, where the star formation is likely to be occurring, are always opaque to visible radiation.

Two surprising discoveries have been made in the last decade or so about HI and HII regions.

The first discovery is the detection by radio observations of complex molecules in some HI regions. Molecules other than H_2 are generally rare in interstellar space because of the low probability of atoms encountering each other, and because of the difficulty of reactions occurring between them when they do. Yet in some HI regions molecules containing up to 13 atoms have been found. The presence of these molecules led to the coining of the name giant molecular clouds (GMCs) for such regions, as previously mentioned.

The second discovery is of intense maser† emission near some small HII regions. This is again in the radio region and comes mostly from OH (an unstable molecule on Earth) and H_2O molecules.

Having set the scene, let us now return to looking at how stars could form inside GMCs etc. The optical astronomer is at a considerable disadvantage in such a study, because the dust completely obscures the process until long after the star has formed. Until recently, therefore, much of the observational work in this area had to be undertaken by radio astronomers whose longer wavelength operations are little affected by either the gas or dust.

Several major advances however have been made in the last two or three years as a result of the IRAS‡ spacecraft's work. This detected and gave accurate positions for hundreds of 'hot spots' or strong infrared sources within GMCs: now thought to be stars in the process of formation (protostars).

Let us start then with a reasonably typical, large HI region. Its average density may be about 10^9 atoms and molecules per cubic metre, or a thousand times the density of the interstellar medium. Its temperature will be about 100 K. These averages though, are likely to obscure wide variations, and in places we may expect the density to be ten or more times greater or smaller than this figure. It may be that such density fluctuations by themselves are sufficient to initiate the next stage of the process. Indeed, for the very first star to form in the universe that surely must have been the case.

Generally, however, it is thought that a 'trigger' of some sort is required to initiate star formation. The trigger acts to push the density and hence mass within some portions of the whole cloud over a limit, known as the Jeans' mass. Once its mass rises above the Jeans' mass, the portion of the cloud concerned will start to collapse under its own gravitational self-attraction. Possible trigger mechanisms could include

---

† A maser is the longer wavelength version of the laser. The name is an acronym standing for **m**icrowave **a**mplification by **s**timulated **e**mission of **r**adiation. It is easier to get stimulated emission at longer wavelengths and so the successful construction of masers on Earth predates that of lasers, though the latter have become more familiar from applications such as compact disc players etc. Similarly it is easier for the conditions for maser emission to occur in nature than would be the case for laser emission. Nonetheless, the conditions for maser emission do require an unlikely combination of circumstances, and so their discovery near HII regions came as a surprise.

‡ The **i**nfrared **a**stronomy **s**atellite. This was launched in 1983 and provided a detailed and precise survey of the sky at long infrared wavelengths. The telescope was cooled by liquid helium to reduce the background noise, and functioned for ten months until the helium ran out.

collisions between clouds, pressure from nearby expanding HII regions or supernova remnants (Journey 7), or a density wave spreading out from the centre of the galaxy following a Seyfert-type explosion (Journey 9).

Now the Jeans' mass depends on both the temperature and the density of the material. For the values typical of the nebula as a whole the Jeans' mass would be about 10 000 solar masses. If a small region reached a density of 10^{11} particles per cubic metre, one hundred times the average value, then the Jeans' mass would fall to 1 000 solar masses. However, the temperature in such a region would also be lower than the average, perhaps only 50 K. This is because the material loses energy mostly via emissions from rotating molecules. The rate of such emissions increases as the square of the density of the material, because the collisions which cause the molecules to rotate become more frequent as the density rises. At 50 K and 10^{11} particles per cubic metre then, the Jeans' mass would fall to less than 500 solar masses and would be contained within a region about one light year across.

The dense region would be emitting less energy than the rest of the nebula because of its lower temperature, and the dust within it would absorb radiation strongly. We might therefore expect to observe some of these regions silhouetted against the background stars as small dark absorption nebulae. It is possible, though by no means certain, that Bok globules (figure 3.4) are at just this stage.

Once such a dense region has formed, it will start to collapse under its own gravitational forces, taking 100 000 years or so to halve in size. As the region collapses, however, its density will increase and its temperature will fall further.

At 10 K and 10^{14} particles per cubic metre, the Jeans' mass will reduce to one solar mass. The collapsing cloud will thus become unstable. It will tend to fragment into numerous smaller cloudlets a tenth to a hundredth of a light year in size, each with a mass comparable with that of the Sun, and each collapsing individually in its turn. Such fragmentation is likely to be assisted by further shock waves from supernovae, HII regions, etc.

We may now concentrate on one of the cloudlets. We can reasonably call the cloudlet a protostar by this stage. Other than through the occurrence of a nearby supernova, or of something else equally dis-ruptive, the cloudlet is likely to condense down to a single star or pair of mutually orbiting stars (binary system).

The protostar will continue to collapse, taking perhaps 10 000 to 100 000 years to reach a size of 100 to 200 AU†: comparable with the

---

† The astronomical unit. This is a convenient unit to use within the solar system or for similarly sized objects. It is the mean distance between the Earth and the Sun: 1 AU $= 1.5 \times 10^8$ km $= 1.5 \times 10^{-5}$ ly.

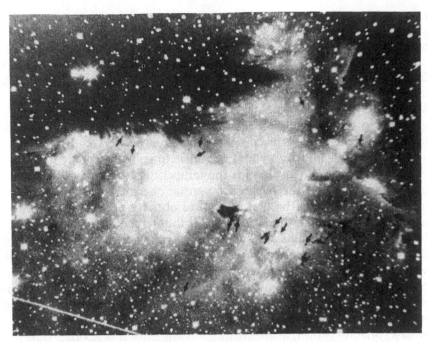

Figure 3.4   Bok globules (arrowed) in NGC 281. (Reproduced by permission of the Observatories of the Carnegie Institute of Washington.)

present size of the solar system. The density within the protostar will then be in the region of 10^{18} to 10^{19} atoms and molecules per cubic metre. With the temperature of the material still at 10 K or lower, the Jeans' mass may fall to less than a thousandth of a solar mass. Further fragmentation within the still-collapsing protostar may then occur and provide condensations which later evolve into planets etc. Alternatively it may split into larger units and so produce a close binary or multiple system of stars.

This point in the collapse is critical for another reason as well as perhaps being the point at which protoplanets come into being. At 10^{18} to 10^{19} particles per cubic metre, the material becomes opaque to the molecular rotation emissions that have so far kept the temperature down. However, as the material collapses, large amounts of potential energy are being released—probably some 10^{38} J by this stage, or the amount of energy that the present Sun would require 10 000 years to radiate. This energy has previously been lost via the molecular line emissions, but now as the material becomes opaque it is trapped inside the collapsing protostar.

The temperature, at least in the inner parts of the protostar, starts to rise rapidly as a result of the continuing potential energy release. The rising temperature in turn leads to rising gas pressures, which are soon sufficient to halt the free fall within the central core of the protostar. The core then contracts much more slowly as it accumulates more material from the outer parts of the protostellar cloud. This slow collapse could conceivably continue and lead to the eventual formation of stars, but it seems likely that this would take longer than the present age of the galaxy. So such a collapse cannot provide the mechanism for the formation of the myriads of stars that we actually observe.

The protostar, however, could continue to collapse rapidly if another 'sink', other than the increase in the temperature of the material, could be found to take up the released potential energy. That sink is indeed found in most cases. The first of several of these occurs as the central temperature rises to around 2 000 K. At that temperature (hotter than a blast furnace, so conditions have changed very radically in a very short time) the hydrogen molecules which make up the bulk of the material split (dissociate) into hydrogen atoms. This requires about 2×10^8 J for each kilogram of hydrogen, or about 4×10^{38} J for the whole protostar. So long as the pressure has not risen sufficiently to bring things to a halt by this stage, the collapse can continue, at an ever-increasing rate, and with a constant central temperature for the protostar of about 2 000 K.

Although the protostar contains a large amount of molecular hydrogen, it is a finite amount, and eventually it will all have dissociated, at least in the core. The central temperature will then start to rise again. At 5 000 to 6 000 K, a second energy 'sink' becomes available: the ionisation of hydrogen. This will mop up another 3×10^{39} J, before the bulk of the hydrogen has been split into free protons and electrons.

Hydrogen can only be ionised once, so it cannot provide any further energy sinks. However, helium is also present in the protostar to the extent of about one atom in ten, and it can be ionised twice, requiring a total of 10^{10} J kg^{-1}, or about 2×10^{39} J for the whole protostar.

These successive energy sinks allow the collapse to proceed until the protostar is reduced to roughly stellar dimensions.

After the second ionisation of helium there are no significant further energy sinks, and the central temperature is free to shoot up until the collapse is halted by the increasing gas pressure. It is likely that this pressure equilibrium will not be achieved until the central temperature is 10^6 K or more. Before such a temperature is reached, nuclear reactions involving lithium and deuterium (heavy hydrogen) will have commenced. These will push the central temperature up towards the 10^7 K required for the conversion of hydrogen to helium (Journey 4). The start of these normal nucleosynthetic reactions, for most people, is the transition point between a protostar and a true star.

Thus in one sense we have reached the end of our journey, having traced the development of a protostar from GMC to star. However, we have concentrated almost exclusively on what is happening deep inside the core of the protostar. Little of what has just been described would be visible to astronomers, whatever wavelengths they might be using. Thus we must return to an earlier stage and look at what is likely to happen to the outer parts of the protostar, in order to interpret the observations.

The collapse of the nebula does not proceed uniformly. The inner parts contract much more rapidly than the outer portions. Thus when the collapse of the protostar's core has been halted, there is still material raining down onto it from further out. This has the effect of increasing the mass of the protostar. The mass increase in turn then leads to rises in the central temperature, pressure and density, and so encourages the start of nucleosynthesis.

However, there is a second effect of this accretion of material which is more immediately observable. As the material collides with the surface of the protostar, its kinetic energy is converted into heat and, as we have already seen, a large amount of potential energy is released by such infalling material.

Of course, the material does not generally fall directly onto the surface of the protostar. The original cloudlet is likely to have had some small rotational velocity, if only because in its orbit around the galaxy the outermost side of the cloudlet would try to move at a different rate from the side closer to the galactic centre. Such rotational energy, or angular momentum, is conserved as the cloudlet collapses. Thus, just like a spinning ballerina speeding up as she brings her outspread arms in towards her sides (figure 3.5), so the conservation of angular momentum

Figure 3.5   Conservation of angular momentum in a spinning ballerina.

will require the collapsing cloudlet to rotate more quickly as it contracts. Thus by the time the protostar has formed it will be spinning rapidly. The subsequently infalling material, having come from further out in the cloudlet originally, and therefore having a greater angular momentum per unit mass, will go into orbit around the core, forming an accretion disc (figure 3.6).

The angular momentum of the material will subsequently be redistributed into turbulent eddies within the disc, and the material will therefore eventually rain down onto the surface of the protostar's core. The

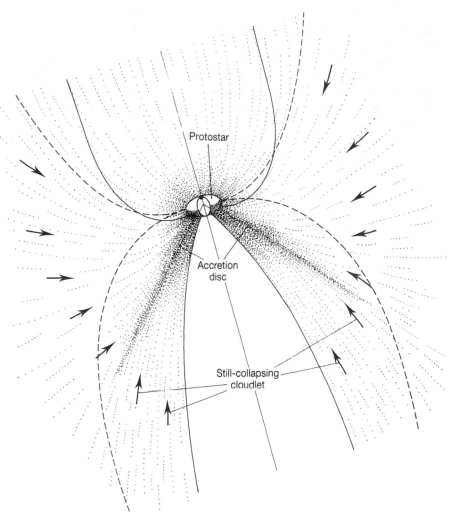

Figure 3.6   The structure of the protostar, accretion disc and cloudlet.

net effect is just to slow down the release of the potential energy, and not to halt it completely.

The protostar, and eventually the star itself, is left rotating rapidly after the contraction phase. The rotation is in fact too rapid to fit in with the observed rotational velocities of the smaller stars and so, as we shall see later, some process must be invoked to slow down the protostar.

At this moment, therefore, the outer structure will comprise the core of the rapidly rotating protostar, with a surface temperature of some thousands of degrees, the accretion disc at about 400 K, stretching out to 50 or 100 AU in the equatorial plane, and the whole still surrounded by a collapsing nebula at 25 to 50 K and perhaps up to a tenth of a light year across (7 000 AU) (figure 3.7).

The differing temperatures of the components of the system at this stage result in their emissions occurring primarily within different wavelength regions. Thus the core has a similar surface temperature to the present Sun, and so will radiate energy mostly over the optical range (100 nm to 1 micron). Although the core is very luminous (the Sun's

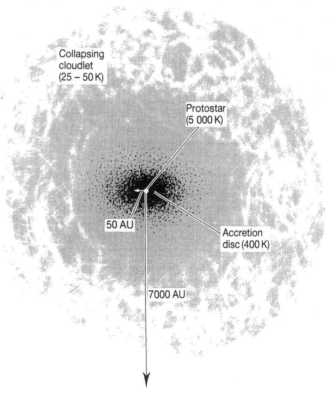

Figure 3.7   The outer parts of the protostar.

core, when passing through this stage, would have been about 10 times brighter than the entire Sun is at present), almost all its energy will be absorbed within the accretion disc, the outer still-collapsing portions of the cloudlet, or within the surrounding GMC. Thus the core cannot be directly observed.

The significance of the IRAS spacecraft observations comes from their extending to 100 microns and so covering the wavelengths at which the accretion disc emits most of its energy. The rest of the cloudlet and GMC are almost transparent to such long-wave infrared observations, so IRAS allowed the accretion discs to be observed directly. Since the accretion disc formation is only a brief phase in the life of a star†, we may expect there to be very few stars actually at this stage at any one time. Indeed, IRAS succeeded in finding only one with reasonable certainty. This is the infrared source in Taurus (figure 3.8) known as IRAS 04365 + 2535‡. Though its core has as yet only a fifth of the mass of the Sun, the object's (infrared) luminosity is currently three times the total solar luminosity. The low temperature outer portions of the protostar will still be observable in the radio region from their molecular emissions.

Now we might reasonably expect that the final star would simply form as the remaining material of the cloudlet accreted onto the core. However, the actual process seems to be much more complex, and during the later stages material is expelled from the system rather than being accreted.

The reasons why an inflow of material changes into an outflow are unclear, although several possible explanations have been suggested. Nonetheless, such a change must occur somehow, for there are at least two strong lines of observational evidence for expulsion of material during the later stages of star formation.

---

† In a stable, random aggregation of a large number of objects, each of which goes through the same series of phases, the proportion to be found at any given instant in a given phase will be proportional to the lifetime of that phase: the longer the lifetime, the greater the number of objects. In effect, the phase with the longest lifetime acts as a 'valve' to regulate the overall rate of progression. Thus in the production of whisky or brandy the bulk of the product is maturing in casks since that stage can last many years. Comparatively small amounts will be fermenting, and only a tiny fraction being distilled (the shortest stage). Another more astronomical examples comes from the generation of energy within stars by the CNO cycle (Journey 4). The slowest reaction (longest lifetime) is the combination of a proton with a nitrogen-14 nucleus, and so nitrogen is by far the most abundant of the three intermediate elements (carbon, nitrogen and oxygen).

‡ The numbers give the approximate position of the source. In this case it may be found near right ascension 04h 36.5m, declination + 25° 35'.

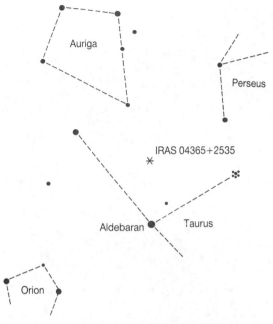

Figure 3.8   IRAS 04365 + 2535.

The first of these lines of evidence is the direct observation of the outflows via their molecular emissions in the radio region. The direction and magnitude of any velocities along the line of sight can be found from the Doppler shifts (Journey 2) of the wavelengths of any emission or absorption lines being produced within the moving material. Thus when most of the sources suggested as possible protostars were observed in detail using microwave emissions from carbon monoxide, the cold gas was found to be moving rapidly away from the core. In most cases, however, the expulsion of the material does not occur uniformly, but is concentrated into two opposing jets. From their structure, such systems have become known as bipolar outflows.

The effects of bipolar outflows can also sometimes be observed in the optical region, when the interaction of the outflow with stationary or slowly moving material well away from the protostar seems to produce small luminous patches termed Herbig–Haro objects (figure 3.9). Some of the maser emissions, particularly from the water (H_2O) molecule, also seem to associate with bipolar outflow interactions with other material.

A straightforward (which does not necessarily imply correct) explanation for the bipolar nature of the outflows could come from the likely structure of the system. Conservation of angular momentum, as already seen, will cause the formation of an accretion disc in the equatorial

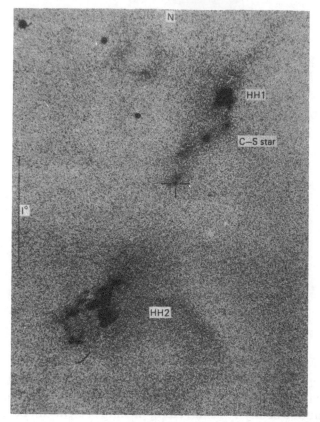

Figure 3.9   Herbig-Haro objects (HH1 and HH2) near the Cohen–
Schwarz star. (Reproduced by permission of the Lick Observatory.)

(rotational) plane of the system. The still-collapsing outer parts of the
cloudlet will also tend to have an equatorial thickening for a similar
reason. The infall of material from above the polar regions, however, will
not have been affected by rotation, and will have accreted much more
completely onto the protostar. Thus even if the initial expulsion of
material were to be isotropic, its flow would be hindered and perhaps
halted by the equatorial concentration of the surrounding accretion disc
and cloudlet, while being relatively free to escape along the rotation axis
(figure 3.10).

The second observational line of evidence for outflow of material from
protostars is less direct, but it is at least as convincing. We have already
encountered it briefly and it lies in the observed rotational velocities of
stars. Doppler shifts can provide us with a means for estimating stars'
rotational velocities. Unless we happen to look directly down onto one of

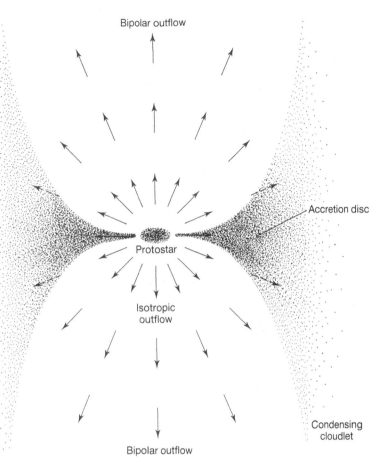

Figure 3.10   Production of a bipolar outflow by the equatorial restriction of an isotropic outflow.

the poles of a star, rotation will cause one edge (limb) of the star to approach us while the other moves away (figure 3.11). With the Sun we can observe each limb separately, and we thus obtain its observed equatorial rotational speed of about 2 km s^{-1}. Other stars cannot be resolved in this fashion. Their rotation, however, does result in a broadening of their spectrum lines in a particular manner, so that lower limits for their rotation velocities can still be estimated.

The observations (figure 3.12) show that the average equatorial rotational speeds are about 150 km s^{-1} for the hotter, more massive stars, but fall to only a few km s^{-1} for the cooler, smaller stars like the Sun.

Now, if all stars form via the sort of processes at which we have just been taking a look, then we have encountered no reason why the smaller

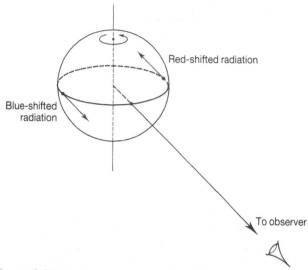

Figure 3.11   Determination of the solar rotational velocity.

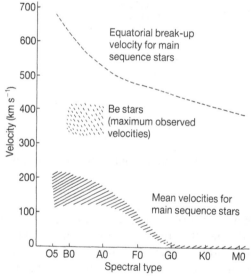

Figure 3.12   Equatorial surface rotational velocities of stars.

stars should have so much less angular momentum than the larger stars. Furthermore, *all* stars would be expected to have significantly higher rotational speeds than those observed, if their angular momentum were conserved from the average rotation and turbulence of gaseous nebulae.

So in order to explain the observations there must be some process acting to slow down stellar rotation, and that process must become much more efficient for stars of less than about one and a half solar masses. Expulsion of material could be the mechanism of just such a process, providing that the expelled material carried away more than its share of angular momentum; then the material left behind would have less angular momentum per unit mass. In other words, the protostar's rotation would slow down.

An apparently complicating factor in this process is that a protostar which is rotating close to its break-up velocity will be enabled to contract further as it loses angular momentum. Its rotation will then actually speed up. Thus the loss of angular momentum must continue long enough for the gravitational and pressure forces to balance each other at the equator as well as near the poles, before the star will actually slow down.

Whether the expulsion of material slows down the protostar's rotation or not, explaining the actual expulsion itself remains a problem at the time of writing. Several possibilities have been suggested, of which radiation pressure is probably the leading contender.

Radiation pressure is not a force with which we are normally familiar; but electromagnetic radiation (light, radio waves, ultraviolet, etc) does exert a force on any object absorbing or reflecting it. This force is normally very small—for someone lying on the beach sunbathing with the Sun overhead, it would be about 10^{-8} of the gravitational force—but its existence can be demonstrated by careful measurement or by the use of a Crooke's radiometer (figure 3.13).

Near to stars, especially the brighter ones, the radiation intensity can become so high that radiation pressure does become a significant force. Indeed, for the largest stars (a hundred or more times the mass of the Sun) radiation pressure can threaten to blow them apart. Many other stars are known to have strong outflows of material, called stellar winds, produced by the pressure of radiation. These winds can reach velocities of $2\,000\ \text{km s}^{-1}$ or more and carry off as much as a solar mass every few thousand years or so. Such winds, however, are generally the result of the pressure of short-wave ultraviolet light acting on atoms in the star's atmosphere, and so occur only around the hottest stars.

Now, although the core of the protostar is bright it is not necessarily very hot, and so it will emit little ultraviolet light by which to power an outflow of this hypersonic stellar wind type. However, an alternative mechanism is available via the dust particles.

As mentioned earlier, the original HI region would have contained about 1% of its mass in the form of small dust particles. These would probably have a core of graphite or silicate minerals a few tens of nanometres across, and would be surrounded by a mantle of frozen gases such as methane, ammonia and water. The particles would probably be

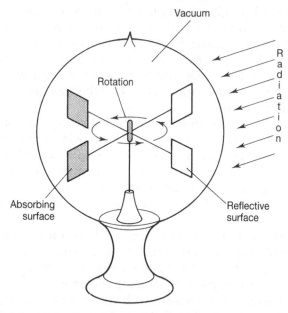

Figure 3.13   A Crooke's radiometer. The vanes are reflective on one side and absorptive on the other. Twice the radiation pressure is exerted on the reflective sides because of the incident and reflected beams, compared with the absorptive sides where there is only an incident beam. Thus the vanes rotate as shown. The cheap versions of the device sold in joke shops have too high a gas pressure inside and rotate in the other direction because the gas molecules bounce off faster from the hotter dark surfaces, than the cooler reflective surfaces.

needle-shaped, with a long axis of a few hundred nanometres or so. During the collapse of the protostar, the icy mantles of the dust particles would be likely to evaporate, but their refractory cores would generally remain in existence. These remnants of the particles of the interstellar dust will absorb radiation from the core of the protostar very efficiently. Since the radiation will be coming from only one direction, but will be re-emitted in all directions, the particles will experience a net force pushing them outwards. Collisions of the particles with the surrounding gas molecules will then transfer the outward force and motion to the material as a whole, resulting in an outflow from the protostar.

Though radiation pressure on dust particles is an attractive explanation for the outflows from protostars, it does have weaknesses. In particular, it is not clear that it can provide sufficient energy for the observed bipolar outflows, or slow down the stars' rotations sufficiently, nor does it explain why the smaller stars should be preferentially slowed.

Alternative explanations for the outflows could come from the operation of magnetic, tidal, convective or turbulent forces, or from shock waves in the accretion disc. Any of these could provide a mechanism for the transfer of angular momentum from the protostar to the accretion disc, slowing the rotation of the star and driving the disc outwards. Possibly a combination of such processes together with a radiation pressure-driven stellar wind may prove to be the correct explanation.

If some such process of angular momentum transfer is involved, then the faster rotation of the larger stars could arise naturally if their stellar winds were disproportionately stronger than those of the smaller stars (as may well be the case). Thus there would be less time for the transfer of angular momentum before the material surrounding the larger protostars became completely dispersed.

Further evidence supporting a form of momentum transfer comes from taking a closer look at the solar system. Not only does the Sun rotate very slowly, but it contains as little as about 2% of the angular momentum of the solar system. The remaining 98% is stored in the planetary orbital motions. Yet the planets contain only a small fraction of one per cent of the mass of the solar system. Such a disparate distribution of angular momentum with mass could have arisen only if some of the Sun's angular momentum had been transferred to the planets.

Since the flow of material, for whatever reason, changes from an inflow to an outflow, there must be a transition stage. During this interval we might see the bipolar outflow of material combined with a continuing infall around the equatorial regions. It might well be expected that such a transition would last only a short while, and so there would be few protostars in this stage at any given time. Nonetheless, IRAS did find exactly such a system: IRAS 16293−2422. This is about 500 ly away in the southern constellation of Ophiuchus, close to Antares (figure 3.14), and radio observations detect a cold infall of material, whilst simultaneously the infrared observations show hot gas being expelled.

It is not clear at what stage planets would be likely to form within the collapsing protostar and cloudlet. Stable condensations may well form at an early stage when the Jeans' mass drops to a small fraction of a solar mass soon after the start of the collapse. These may then collide to form larger condensations within which solid particles may grow around the nuclei provided by the interstellar dust grains. Further collapse may then lead to solid bodies with diameters in the range of metres to hundreds of kilometres.

It may seem obvious that the basic structure of the solar system—the terrestrial planets formed of refractory minerals, and the jovian planets formed of volatile gases—would arise via a temperature gradient within the nebula surrounding the protostar. The material closer to the

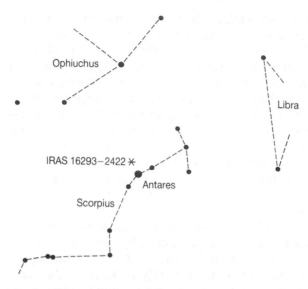

Figure 3.14   IRAS 16293 − 2422.

protostar would be expected to be hotter than that further away, thus encouraging the condensation of volatile material in the outer reaches of the nebula. However, the temperature of 400 K or so measured for the accretion discs may be too low for this apparently obvious process to operate.

An alternative explanation for the way in which the division of the solar system into the terrestrial and jovian planets arose might be provided through the dispersal of the accretion disc by the outflow of material from the star. We would expect the refractory minerals to condense into solid particles throughout the disc, and perhaps to accumulate to form larger bodies. Close to the protostar, however, the outflow of material would sweep up the remaining gas before the planetesimals could grow massive enough to trap it gravitationally, leaving only enough material to form the terrestrial planets. Further out, the accretion disc might last long enough for the small rocky bodies to grow sufficiently large to garner the surrounding gas. They would then grow very rapidly, like a snowball rolling down a hill, since there would be so much material available in the form of the gases. The rocky nucleus would thus quickly be buried under a massive accumulation of the gases, and so the giant jovian planets would originate.

Whatever the precise process of planetary formation, it seems likely to be an intrinsic part of star formation. We should thus expect planetary systems, such as the solar system, to be very common, especially amongst the smaller, very slowly rotating stars. Such stars form 90% or more of the stars in the galaxy. Apart perhaps from stars in very close binary systems whose planets would be likely to be lost or destroyed, we can therefore expect most stars to have planetary systems.

Evidence to support the plurality of planetary systems comes from several sources. Some nearby stars have had their positions and velocities measured sufficiently accurately to show that they are moving in small ellipses. This is a common phenomenon and implies that the star is actually a member of a binary system and is orbiting the common centre of mass. However, in some cases the mass of the companion turns out to be only a few times that of Jupiter, far too small for it to be a star, so the companion must be a large planet. The second line of evidence is less direct and is that the IRAS spacecraft has found discs of dust particles surrounding the otherwise normal stars, Vega and Beta Pisces. Such discs are not necessarily indicative of the existence of planetary systems, but since the solar system has a thinner version of just such a dust disc in the plane of the ecliptic, it could be a hopeful sign.

Returning to the system as a whole, the outflow of material will disperse the accretion disc and surrounding cloudlet, probably in less than a million years. The majority of the infrared sources found embedded within GMCs are shown by radio observations to be at just this stage. The accretion discs themselves absorb much of the energy coming from the protostar, and re-radiate it at short infrared wavelengths.

Observations, however, show that the discs are too bright for this to be their only source of energy. Potential energy is thus presumably still being released as outer parts of the cloudlet continue to fall inwards, even though the outflow from the protostar is already sweeping up the inner parts.

At some stage, the protostar must also start the nucleosynthetic energy-generating reactions in its interior. At one time this was thought to be a traumatic process which would change the star's structure radically. Now it appears likely to pass unnoticed since the nuclear fusion reactions release much less energy than that still coming from the infalling material.

The majority of stars, over 90% by number, are similar to the Sun in their basic structure and belong to a class known as the main sequence. This term comes from a common way of displaying the main physical characteristics of stars. A scattergram is plotted with the star's tempera-ture along the horizontal axis, and its luminosity along the vertical axis. The resulting picture is known as the Hertzsprung–Russell diagram (figure 3.15) after the two astrophysicists who started much of the work in this area at the beginning of the century. For historical reasons the temperature is plotted as increasing towards the left. As can be seen, most of the stars fall within a narrow curving band from the top left to the bottom right on this diagram. This region is called the main sequence,

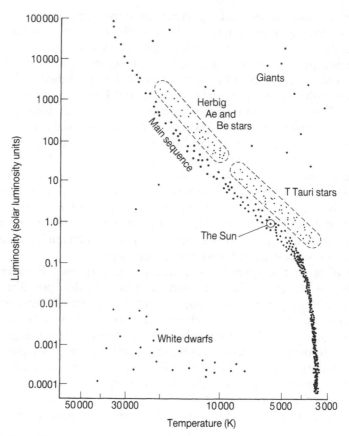

Figure 3.15   The Hertzsprung–Russell diagram for nearby stars, showing the T Tauri and Herbig Ae and Be stars. The relative numbers of the latter types of stars have been much exaggerated for clarity.

and stars within it are therefore known as main sequence (or dwarf) stars.

Most of the remaining stars fall within a less well defined region at the bottom left of the diagram. They are the white dwarfs which we shall encounter again in Journey 5. That leaves about 1% of the stars unaccounted for, and the bulk of these are scattered over the top right-hand side of the diagram and are the various forms of giant star.

A very small number of stars, less than 0.1%, do not fall into any of these main classes, and a large variety of often extremely interesting classes of variable or otherwise peculiar stars are found amongst them. One of these subsidiary classes is known as the T Tauri stars, after the first of the type to be noticed. Their temperature and luminosity place these T Tauri stars in a band just above the main sequence (figure 3.15), and they probably join with another group known as the Herbig Ae and Be stars at the higher temperatures.

About 1 000 T Tauri stars are known and one of their main characteristics is a close relationship with HI and HII regions. They often form groups known as T associations within such gas clouds. T Tauri stars' other major properties are an irregular variability, emission lines in their spectra, large infrared emissions and an excess of light elements, especially lithium, compared with the Sun and most other stars.

Long before detailed theories of the origins of stars had been worked out, the T Tauri stars were thought to be very young stars. The reasons for this were twofold. Firstly, there was their association with gas clouds where stars were likely to be forming; but of more significance was the lithium excess. This is because the lithium is probably produced in the interstellar gas by collisions with cosmic rays†, although it is destroyed easily and at comparatively low temperatures by nuclear fusion reactions (hence its use in so-called 'hydrogen' bombs). Thus once such reactions start inside stars, the lithium will disappear in a million years or so, and any star containing an excess of lithium cannot have had nuclear reactions occurring within it for any length of time.

The final stage in the formation of a star therefore almost certainly occurs as it passes through the T Tauri region. It will still be embedded in a thick envelope of material, mostly composed of dust and gas but perhaps also with planets or protoplanets in the process of formation. Any remaining gas will be being rapidly expelled at velocities of several hundred kilometres per second. The star itself may be a variable, or its brightness may vary as it is obscured now and then by thicker portions of the surrounding envelope.

Over the next few hundred thousand years the star will decrease in brightness and increase in temperature slightly (implying a continuing decrease in size), until it joins the main sequence. Then, depending upon its mass, it will settle down to a quiet 'middle age' lasting between 10^7 and 10^{12} years. Later, the star may go on to a spectacular old age, but

---

† Cosmic rays are not radiation at all, but sub-atomic particles travelling at close to the speed of light. Most of the particles are protons or helium nuclei, with a few electrons and very small numbers of heavier nuclei. They are currently thought to be produced in supernova remnants (Journey 7). Their extremely high energies result in many reactions occurring should they collide with other particles. Thus the light elements such as boron, beryllium and lithium can be produced by collisions between cosmic rays and heavy nuclei in the interstellar medium. The radioactive nucleus, Carbon-14, used by archaeologists for dating their finds, results from collisions between cosmic rays and nitrogen nuclei high in the Earth's atmosphere.

that is another story, parts of which we will encounter on several of our future journeys.

We have not quite reached the end of this journey, however, even though we have now followed through the formation of a star both from the inside and the outside, for that star still lies embedded within the original HI region or GMC. Many stars are likely to be forming nearby and nearly simultaneously, and before the GMC completely disperses a cluster of gravitationally linked stars is likely to result (figure 3.16). Such clusters are called galactic clusters and can contain from a few to a few hundred individual stars. In the cluster's orbit around the galaxy, those stars on its outer edge will be trying to move at different velocities from those nearer the galactic centre. Over the next hundred million years or so, the cluster will therefore split into single stars, close binaries and multiple systems.

The larger, more massive stars will have other effects before the cluster disperses. Firstly, they will form HII regions within the larger HI region. This occurs because the larger stars are also the hotter stars, and they emit copious amounts of ultraviolet radiation. Those photons with wavelengths less than 91.2 nm have sufficient energy to ionise the hydrogen atoms of the surrounding gas cloud, and so a bubble of hot ionised hydrogen forms around the star: a nascent HII region.

(a)

(b)

Figure 3.16   Galactic clusters. (a) Pleiades. (Reproduced by permission of the Palomar Observatory.) (b) h and χ Persei. (Reproduced by permission of the Lick Observatory.)

Infrared radiation re-emitted from the HII region may then spark off the OH and H_2O masers found in some GMCs. The size of the HII region increases until the number of ionisations just balances the number of electrons and protons recombining to form hydrogen atoms, incidently wiping out the masers on the way. The recombinations often result in the emission of visible radiation, and so we eventually see the HII region as a glowing nebula within the larger, cooler cloud.

The HII region around a single star formed within a uniform HI region would be a perfect sphere. However, several HII regions from different stars often merge, and the HI region is also likely to be very variable in its density. Thus we get spectacular and often irregular nebula shapes such as those shown in figure 3.2.

The gas in the HII region, as well as the region itself, will be expanding because it has been heated by the star(s) to some 10 000 K. The internal pressure in the HII region is thus much greater than that in the cooler (100 K) enveloping HI region. The expanding HII region will compress the gas of the Hl region at their interface. Perhaps then the cycle may be completed through the initiation of a new phase of star formation in this compressed zone. Eventually, however, the expansion will disperse the remaining gas back to the general interstellar medium.

In a few million years, the largest stars will complete their lives and explode as supernovae (Journey 7). In many cases this may occur whilst they are still embedded within the nebula. Then the resulting much more violent shock waves may trigger star formation just as do the expanding HII regions.

The supernovae also have another, longer term effect. When they explode, the star is almost completely disrupted, and the products of the nuclear reactions in its interior are distributed back out into the interstellar medium. In some types of supernovae these products may include all the familiar 92 elements, and some of the transuranian elements and/or isotopes no longer naturally in existence on the Earth.

Thus as time goes by, the interstellar medium must change its composition from the initial almost pure hydrogen and helium, to become richer in the heavier elements. This change can be observed, and helps to assign ages to stars. Thus the stars in the nucleus of the galaxy (Journey 9) and in the globular clusters, thought to be 8 to 12 aeons old, contain only about 0.5% by mass of elements heavier than hydrogen and helium. The Sun, formed 4.5 aeons ago contains about 2%, and stars forming today, about 4% of such elements.

A supernova may also perhaps explain a peculiarity of the solar system. The equatorial plane of the Sun is inclined at about 6° to the mean plane of the orbits of the planets. Now as we have seen, the collapse of the protostar should leave an equatorial disc within which the planets would form, and which should lie very close to the rotational plane of the protostar. Only a nearby supernova occurring in the later stages of the formation of the Sun would be likely to inject enough material into the accretion disc and at a sufficiently high angle to change its orbital plane by as much as the observed 6°.

Such a discrepancy between the rotation and orbital planes could be dismissed as a random fluke, were it not for another strand of evidence for a supernova explosion occurring during the formation of the solar system. That evidence comes from the meteorites. A 10% excess of the isotope magnesium-26 is found in the Allende meteorite, which is thought to have been little changed since the origin of the solar system. This isotope is a decay product of the radioactive isotope, aluminium-26. Since the latter has a half-life of only 720 000 years, none of it exists naturally now within the solar system. However, aluminium-26 would be expected to be found in the ejecta from a supernova. Other similar isotope anomalies are also found, suggesting the presence of debris from a supernova. Thus it would seem likely that this meteorite, and hence presumably the rest of the solar system, must have been formed partially from the ejecta from a supernova within at most one or two million years of its explosion.

The births of stars, and more particularly the birth of the solar system, are thus intimately connected with the deaths of stars, the one rising phoenix-like from the ashes of the other. Numerous supernovae must have occurred between the formation of the galaxy and that of the solar system in order to build up the heavy elements forming the Earth and ourselves. It is likely that the collapse of the cloud could have been initiated by a pressure wave from a supernova. Finally, as we have just seen, a supernova may well have occurred within a few light years of the nascent Earth and remainder of the planetary system. The presumptious Glendower quoted at the start of our journey could have little known how truly he spoke.

Journey 4

The Nearest Star

A tiny Sun, and it has got
A perfect glory too;
Ten thousand threads and hairs of light,
Make up a glory, gay and bright,
Round that small orb, so blue.

The Three Graves pt IV
S T Coleridge

Great concern is currently being felt over the 'greenhouse effect', the slow increase in the amount of carbon dioxide in the Earth's atmosphere due to the consumption of fossil fuel and reduction in plant biomass, leading to increases in the average surface temperature of the Earth. The effect is exacerbated by the consequent increased release of methane and the evaporation of water as the temperature rises, because all three gases allow through the short-wave radiations from the Sun whilst trapping the long-wave re-radiation back into space by the Earth. Dire predictions of rises in the global average temperature by the middle of the next century are given.

The effects of such an increase, even though of only a few degrees, may well be drastic. Sea levels may rise to wipe out low-lying countries like Bangladesh and the Netherlands and put many cities and ports in danger. The level will rise not because of the often-mentioned melting of the polar caps, but because the sea water will expand in volume as its temperature rises. Climatic patterns will also shift, and agricultural zones will change.

However if the predictions are wrong, and we all know how far ahead even local weather is capable of prediction, then a far worse situation could face us: the 'runaway greenhouse effect'. The released methane and water vapour could trap sufficient heat, without any further increase in the carbon dioxide levels, to be able to release more methane and water vapour. That in turn would raise the temperature further, releasing yet more gases and so on. A positive feedback loop with disastrous consequences for most life on Earth. The final surface temperatures might possibly exceed the boiling point of water.

Yet we cannot be certain that the greenhouse effect will take place so neatly. Many other influences are going to come into play as the temperature rises. For example, the increasing amounts of water vapour in the atmosphere will lead to more cloud cover, and this in turn will reduce the amount of solar radiation reaching the surface of the Earth by reflecting more of it directly back into space. Temperatures could possibly then stabilise at only a little warmer, on average, than at the moment, but we would have many more dreary, cloudy summers and dank winters to 'enjoy'. Also, the Sun seems to be temporarily fading very slightly at the moment, and that will tend to oppose the greenhouse effect.

Nonetheless, and irrespective of mankind's efforts, temperatures on the Earth are going to rise beyond the point at which life as we know it can exist, and that is because the Sun on longer timescales is slowly increasing in both its temperature and luminosity. Luckily for the life insurance companies, that increase in the Sun's radiation will occur over vastly longer times than the effects we have just been discussing: at least hundreds of millions of years. On an even longer timescale of 4 or 5 aeons the Earth itself may be destroyed.

What then are the changes in the Sun that are going to worry our great-to-the-nth grandchildren? And how can we know what will happen to the Earth in several aeons when we cannot even predict the weather a month ahead? For answers to those questions we must start by examining what is happening now inside the Sun.

Just as when we looked at how the solar system came into being during the last journey, so when looking into the future we are aided by the Sun being a fairly normal star. We can therefore look at what is happening now to other, older, stars to get an idea of what is likely to happen to the Sun.

At the moment the Sun is generating its energy by nuclear reactions at and near its centre. The source of the Sun's energy and that of most other stars has been known to be some form of nuclear reaction since the turn of the century. Geologists at that time started to show that the age of the Earth, and so of the solar system, was at least several aeons. All other possible energy sources: chemical reactions, continuing potential energy release by collapse, a rain of meteorites onto the solar surface etc, could only maintain the Sun's energy output for a few tens of millions of years at most. Thus nuclear reactions were all that were left.

It was only in the 1930s, however, that the actual reaction routes were deduced. For main sequence stars (Journey 3) like the Sun, the energy comes from the conversion of hydrogen into helium. There are two sets of reactions for this conversion and their details are shown in figures 4.1 and 4.2. The proton–proton chain dominates the energy generation of smaller stars, whilst the CNO cycle supplies the energy of the more massive stars. The Sun, with a mass of 2×10^{30} kg (330 000 times the mass of the Earth), is larger than the average star (7×10^{29} kg), but it is not a real heavyweight. Most of the Sun's energy therefore comes from the proton–proton chain, with only about 5% from the CNO cycle.

At this point we must needs encounter a major current embarrassment for astrophysicists. There is a long-standing failure of predictions based upon our present model of the Sun to match some fundamental observations: those of the subatomic particles, neutrinos, given off in several of the reactions shown in figures 4.1 and 4.2.

We first encountered neutrinos briefly during Journey 1; now, however, we need to know a little more about them. Their main characterist-

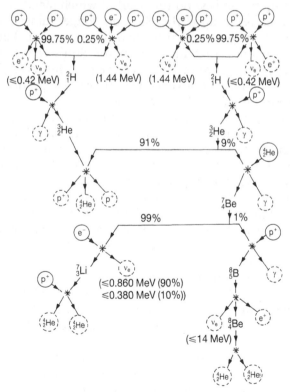

Figure 4.1   The proton–proton chain and its variants. Particles consumed in the reactions are encircled with a solid line, final products with a broken line. Neutrino energies are shown in parentheses.

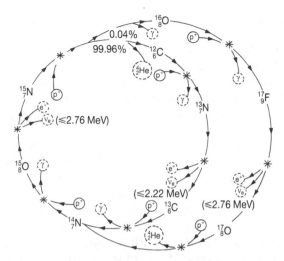

Figure 4.2   The CNO cycle. Particles consumed in the reactions are encircled with a solid line, final products with a broken line. Neutrino energies are shown in parentheses.

ics seem to lie in differing in very few respect from being nothing at all! They have zero electric charge, zero electric moment, zero magnetic moment and zero or very small rest mass. About their only property is spin, which in the units used by nuclear physicists has a value of a half.

Additionally, neutrinos interact very poorly with other forms of matter. The first operational method for observing them relied upon detecting a neutrino combining with a nucleus of the chlorine-37 isotope. Even though this is a relatively strong interaction, neutrinos are so unreactive that a beam of them would have to pass through about 3 ly of pure chlorine-37 before being reduced to half intensity! This very low interaction probability means that neutrinos produced by nucleo-synthetic reactions in the core of the Sun pass through the overlying 700 000 km of material almost without hindrance. We should therefore be able to observe the centre of the Sun directly and with only about an eight minute delay (because neutrinos travel at close to the speed of light) providing only that we have a suitable neutrino 'telescope'. By contrast, if we observe the Sun via light, radio, x-rays etc, then that energy has taken about ten million years to percolate from the centre to the surface, and has been changed many times along the way. It can therefore tell us very little directly about conditions at the centre.

Just such a neutrino 'telescope' has in fact been in operation for some two decades. It is actually just a detector, not a telescope, since no indication of the direction of arrival of the neutrinos can be obtained. The detection relies upon the reaction with chlorine-37 referred to above. In this reaction an argon-37 nucleus is produced which may subsequently be detected as it decays radioactively back to chlorine. In order to catch even a few of the elusive neutrinos, 600 tons of tetrachloroethene (each molecule of which contains one chlorine-37 atom on average) are needed, and the tetrachloroethene has to be buried 1.5 km down a gold mine at Homestake, South Dakota, to shield it from other contaminating reactions.

The results (figure 4.3) have averaged about one neutrino being caught every two and a half days. The predictions (figure 4.4), however, based on current ideas of what is going on inside the Sun, suggest the rate should be about one neutrino every 16 hours. There is thus a discrepancy between prediction and observation of a factor of three.

However, the Homestake detector cannot detect the bulk of the solar neutrinos (figure 4.4), only those with energies greater than 0.814 MeV. It is hoped that new types of detector, based upon gallium-71 and capable of detecting neutrinos with energies down to 0.23 MeV, will shortly be completed. Meanwhile confirmation of the discrepancy, or solar neutrino problem as it is usually known, comes from a third type of detector. That detector, at Kamioka in Japan, detects neutrinos as they collide with electrons in a huge tank of water (see also Journey 7). Its energy limit is very high. Only neutrinos with energies above 9 MeV are detected, but the observed rate is again lower than the predicted rate by about the same factor of three.

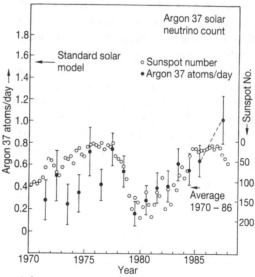

Figure 4.3   Solar neutrino counts from the Homestake detector. (Reproduced by permission of Professor D H Perkins, University of Oxford.)

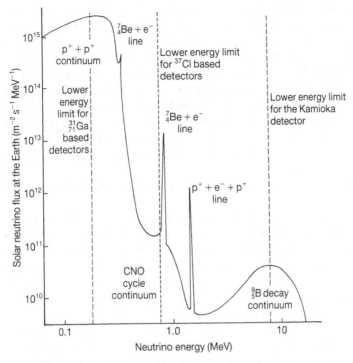

Figure 4.4   The predicted solar neutrino spectrum.

No fully satisfactory explanation of the solar neutrino problem has yet been suggested. If the discrepancy does turn out to apply to all the solar neutrinos, and not just to the high energy ones detectable at present, then it may become necessary to alter our ideas about what goes on inside the Sun and other stars very drastically.

If we are so uncertain of the conditions inside the Sun now, how can we hope to predict what will happen to it in the future? The answer is found in the observations of other stars. We simply find stars that are as similar to the Sun as possible, but older, and see what has happened to them. The solar neutrino problem may mean that we do not understand *why* the changes have occurred, but the observations *do* show what form they will take.

Thus with some confidence but bearing in mind that some of the explanations may change, we may proceed to look at what is currently being predicted for the future of the Sun and solar system.

We saw on the last journey that the initial source of energy of the Sun was the release of potential energy by the infalling material forming the protosun. That energy source of course died away as the collapse came to a halt. At about the same time nuclear reactions would have started up as the central temperature rose to a million degrees or so. Those first nuclear reactions would probably have consumed the lithium and deuterium (heavy hydrogen), since those elements can react at relatively low temperatures. The energy generated would be small compared with that released by the collapse of the protosun, but it would be being released right in the centre of the core. It would push the temperature in the core up towards the 5 to 10 million degrees required for significant energy to be released via the proton–proton chain. That in turn would release sufficient energy to raise the central temperature towards the current value of 15 million degrees. Thereafter, and until the present day, the Sun's energy has come from the conversion of hydrogen into helium via the proton–proton chain with a small contribution from the CNO cycle.

The Sun currently radiates energy at a rate of 4×10^{26} watts (W): sufficient energy in one second to meet mankind's needs for 200 000 years. That energy comes from the conversion of mass into energy. The four hydrogen nuclei which go into making the single helium nucleus (figures 4.1 and 4.2) are 5×10^{-29} kg heavier than the end products. The production of each helium nucleus therefore releases about 26 MeV $(4.7 \times 10^{-12}$ J). A small proportion of this escapes with the neutrinos, but the rest goes into maintaining the Sun's other energy emissions.

Now the energy produced by a single reaction is negligible, but so many reactions are occurring in the core of the Sun that some 6×10^{11} kg of hydrogen are converted into 5.96×10^{11} kg of helium every second. Thus the Sun converts matter into energy at a rate of 4×10^9 kg s^{-1} (four million tons per second!). For comparison, the Hiroshima atomic bomb probably converted a total of 10 g of matter into energy. The Sun's mass is about 2×10^{30} kg, and at its formation about three quarters of that was hydrogen. Thus it would take, even at what seems a profligate rate by human standards, about 10^{11} years for all the hydrogen to be converted in helium. A length of time which is a comfortable 20 times the present age of the Sun and solar system.

Inevitably, however, such a simple-minded calculation of the solar lifetime is inadequate. Most of the material in the Sun will never experience high enough temperatures and pressures to undergo nucleosynthetic reactions. Perhaps only 10% or so of the hydrogen will have been converted into helium before the structure of the Sun will be forced into radical changes. Thus the Sun's lifetime is reduced to about 10^{10} years. Even that, however, would apparently still give us another 5×10^9 years before we need think of emigrating to another planetary system.

Again, though, we still have too simple-minded a picture. The helium produced by nucleosynthesis accumulates in the core of the Sun. At the moment, the composition at the centre has changed from the initial 74%

hydrogen and 24% helium to about 38% hydrogen and 60% helium. This slow increase in the helium content of the core has the effect of making it easier for the energy produced at the centre to leak out to the surface. The Sun will thus slowly increase in brightness with time.

The Sun is currently 50% brighter than when it first joined the main sequence, and it will increase by a factor of two or so over the next 5×10^9 years. That will increase temperatures on the Earth by 50 to 60 K. In a much shorter time, perhaps as 'little' as 500 million years or the time from the first trilobites to ourselves, the greenhouse effect or the runaway greenhouse effect, even without mankind's contributions, will have raised the *surface* temperature to levels intolerable to all current life forms except some of the blue-green algae. Perhaps a few more complex organisms may continue to exist in the depths of the oceans, but even these will eventually die as the oceans evaporate, leaving the Earth a slightly cooler, but equally barren version of the present-day Venus (figure 4.5).

Some of the other planets will also be affected on this same sort of timescale. Mars is likely to lose the remainder of its atmosphere; Venus' surface temperature may rise sufficiently for some rocks to melt giving it permanent lakes or seas of lava. Alternatively, Venus might start to lose significant fractions of its atmosphere and so its surface might even cool down. Mercury and our Moon will just continue alternately to bake and freeze on their bright and dark sides. At this stage (500 million years in the future), the outer planets and their satellites are unlikely to be much affected by the changes in the Sun.

Figure 4.5   The present-day surface of Venus as revealed by Veneras 9 and 10. (Reproduced by permission of the USSR Academy of Sciences.)

Though our descendants, unless they evolve in ways unimaginable to us, will not be around to witness it, the next chapter of the story will occur some 4 aeons or so from now.

The Sun will by then have completely converted the hydrogen at its centre into helium. Apart from the 2% of material in the form of the heavier elements, the core of the Sun will thus be pure helium. The nuclear reactions will therefore die away at the centre of the Sun, but will continue further out in a shell around the core where hydrogen is still available (figure 4.6).

Once the central nuclear reactions have died away, the Sun's core will achieve a uniform temperature (become 'isothermal') throughout its volume, and this temperature will be equal to that of the base of the hydrogen burning shell. Now the pressure must continue to increase towards the centre in order to support the overlying layers of material, but if the temperature is constant then the pressure increase can only occur through increasing the density towards the centre of the core. However, there is a limit, known as the Schönberg–Chandrasekhar limit, to the mass of the outer layers that can be supported in this way. For the

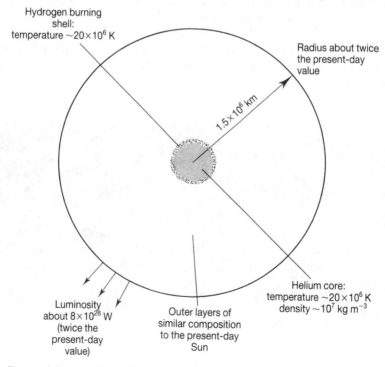

Figure 4.6   A schematic cross section through the Sun towards the end of its main sequence life.

solar-type stars, that limit is reached when the isothermal core reaches about 12% of the mass. Once the core exceeds this mass, it must contract and heat up in order to develop sufficient pressure within itself to support the outer layers.

The core contraction marks the end of the main sequence lifetime, and will occur in about five aeons from now for the Sun. During the interval until the start of the core contraction, the Sun will double in luminosity, and it will increase in size, also roughly doubling over the same time interval. Thus we shall only be able to enjoy the spectacular total solar eclipses (figure 4.7), which rely on the angular size of the Moon being greater than that of the Sun, for about another 100 million years. The resulting increase in the Sun's surface area will lead to a reduction in its surface temperature by 1 000 K to about 5 000 K despite the increase in luminosity. However, since it is the solar luminosity, rather than the surface temperature that determines conditions on the Earth, the fall in solar temperature will not help our descendants.

Towards the end of the Sun's main sequence lifetime, or soon afterwards, the helium core will undergo a fundamental change in its nature. As the density approaches 10^7 kg m^{-3} (10 000 times the density of water, or about 60 times the present central density of the Sun), the material will become electron-degenerate. Such material, and the even denser baryon-degenerate material, is the stuff of white dwarfs and neutron stars, and we shall encounter its peculiar properties many times again (Journeys 5, 6 and 7).

Of importance now is that the pressure exerted by the free electrons in electron-degenerate material is independent of the temperature. The core will therefore be forced to contract with increasing rapidity in order to supply the required pressure by increasing just its density, and this contraction will have two important consequences.

Firstly, potential energy will be released just as it was during the initial collapse from an interstellar gas cloud, and this in turn will raise the temperature of both the core and the surrounding hydrogen burning shell. But the nucleosynthesis reactions depend very sensitively on temperature, and so the amount of energy they produce will rise considerably. The temperature of the Sun's hydrogen burning shell may even rise sufficiently (to over 2×10^7 K) for the CNO cycle to dominate the energy production, instead of the proton–proton chain as at present. The increase in the energy production, of course, leads to an increase in the energy radiated, and so the Sun will start to brighten very considerably.

The second effect of the core contraction is that the outer regions of the Sun will expand. Thus in less than an aeon after leaving the main sequence (i.e. from when core contraction starts), the Sun will be about six times its present size, and twelve times its present luminosity. These changes move it from its present position on the Hertzsprung–Russell

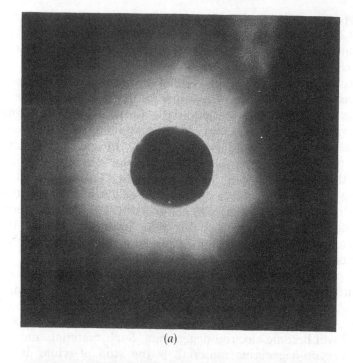

(a)

(b)

Figure 4.7   Total solar eclipses: (a) solar maximum and (b) solar minimum. (Reproduced by permission of the Royal Astronomical Society.)

diagram off towards the upper right, to about position A on figure 4.8.

At this stage, the inner solar system will really start to suffer. Mercury's surface temperature may reach 1 100 K: hot enough to melt most rocks. Venus and Earth are likely to lose their atmospheres and be left with surfaces baked to 600 K and 500 K respectively. Even Mars' surface is likely to reach 460 K, hot enough to melt ordinary solder. The outer planets, however, will still be relatively unaffected, though obviously getting warmer.

The solar core contraction will continue, with the central temperature rising towards 10^8 K. Simultaneously, the outer layers of the Sun will continue to expand and its luminosity to increase while its surface temperature decreases. After a further small fraction of an aeon, just before the next major development, the Sun will have reached 400 times

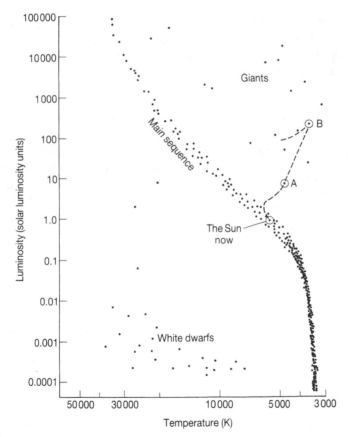

Figure 4.8   The Herzsprung–Russell diagram for nearby stars, showing the evolutionary track of the Sun from the main sequence to the helium flash.

its present luminosity and 60 times its present radius, and its surface temperature will have fallen to 3 500 K.

Mercury will then almost be engulfed by the expanding Sun, and at nearly 3 000 K most of its materials will evaporate. Quite possibly the outer atmosphere of the Sun will cause sufficient drag for the remainder of Mercury to spiral into the Sun and be lost totally. Venus, Earth and Mars may be lost similarly, though on balance they seem likely to survive since the Sun will be at its maximum size for only a relatively brief interval. Their temperatures though are likely to rise to 1 500, 1 200 and 1 100 K respectively, so that they will become completely molten. They will also reduce in mass through the loss of volatile elements. Jupiter and Saturn will reach temperatures of 550 and 400 K, but they are sufficiently massive that even then they will retain most of their gases. Uranus and Neptune will reach a comfortable room temperature and also survive largely unscathed.

The contraction of the core and consequent expansion of the outer layers of the Sun will come to an end when the temperature at the centre rises sufficiently to support helium burning. This will occur at a little over a hundred million degrees, and with the central density approaching 10^9 kg m^{-3}.

The enormous rise in temperature and density required to initiate helium burning (cf 10^7 K and 10^5 kg m^{-3} for the start of hydrogen burning) is a consequence of the large amount of energy required to ignite the reaction and of the instability of the beryllium-8 nucleus. This latter would be the natural product of the fusion of two helium-4 nuclei, but it is highly unstable and fissions back to helium with a half-life of only 2×10^{-16} seconds. Only if a third helium-4 nucleus interacts with the beryllium before the decay, to produce the normal isotope of carbon, can the build-up of elements proceed any further. Significant numbers of what amounts in effect to a triple collision of helium nuclei only start to occur for the densities and temperatures mentioned above. Since the bare helium-4 nucleus is the alpha particle of radioactivity, the helium burning reaction is often known as the triple-alpha reaction. Like hydrogen burning, energy is released because the final product (carbon-12) is less massive than the three helium nuclei combining to form it. A single such reaction though, releases only 7.3 MeV, less than a third of the energy released by the formation of each helium nucleus from hydrogen.

When it finally does occur, the onset of helium burning will be dramatic. The energy released by the reactions will go into heating the centre of the core, but since that is formed of electron-degenerate matter, the rising temperature will have little effect on the pressure and so the structure and density of the core will remain unaffected.

The triple-alpha reaction, however, is extremely sensitive to temperature: doubling it would increase the reaction rates by a factor of 10^{12}! Thus the rising temperature will greatly enhance the energy release, which will then further raise the temperature, and so on. The reactions will start to run away in what is called the helium flash.

A peak in the energy release of 10^{11} times the Sun's present luminosity may be reached for a short time. Such a huge pulse of energy will lift the electron degeneracy, and the consequent sharp increase in the pressure will then cause the core to expand at last. Helium burning is likely to cease as the core expands, resuming later within the reorganised non-degenerate core.

If, as we have seen, energy is released during collapse, then logically it must be supplied during expansion. Thus the energy released during the helium flash will be mopped up by the expansion of the core. There will therefore be little immediate, directly observable effect of the helium flash on the appearance of the Sun at this stage. The expansion of the core, however, will cause the outer layers to shrink, and the Sun will start to reverse its evolutionary track (point B on figure 4.8), becoming fainter, hotter and smaller.

After the helium flash our ideas about what happens to stars become much less clear, but the general nature of what must occur is given by observation. The outer radius of the Sun will continue to shrink, and it will reduce in brightness, though still remaining much brighter than at present. At the same time, its surface temperature will increase.

These changes will cause its position on the Hertzsprung–Russell diagram to move towards the left (figure 4.9). It may have two energy sources during this time: helium burning at the centre and hydrogen burning in a shell further out. Alternatively, hydrogen and helium burning may alternate as the predominating energy sources.

The solar central temperature may rise sufficiently for the carbon produced by the triple alpha reaction to combine with the remaining helium nuclei and produce the most common isotope of oxygen, but this is not certain. The Sun, however, is not massive enough to take nucleosynthesis further (as discussed on Journey 7). The Sun may become unstable and pulsate as a regular variable star with a period of a few hours, though again it may not be massive enough for this to occur. Eventually the changes will bring the Sun to position C on figure 4.9, some 10 times brighter than now and with a surface temperature in the region of 10 000 to 15 000 K. To a visiting space traveller on the remains of the Earth, the Sun would then appear roughly its present size in the sky, but blue in colour.

By the time the Sun has evolved to position C on figure 4.9, some 10^9 years after leaving the main sequence, and perhaps 10^8 years after the helium flash, it will have consumed its available hydrogen and helium and its nuclear reactions will die away. The loss of its main energy

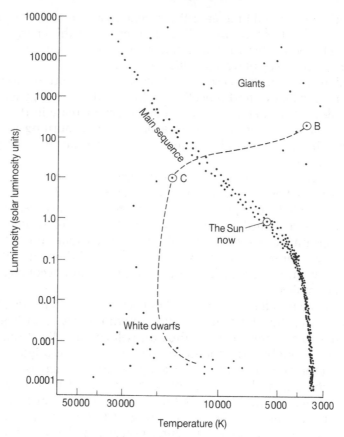

Figure 4.9   The Herzsprung–Russell diagram for nearby stars, showing the post helium flash evolutionary track of the Sun.

sources, somewhat surprisingly, will not cause the Sun to cool down, at least initially. Instead, it will collapse slowly, releasing potential energy, and maintaining its temperature at 12 000 to 15 000 K, or even higher.

These changes will move the Sun's position on the Hertzsprung–Russell diagram vertically downwards towards the white dwarf region (figure 4.9). At various times since leaving the main sequence, the Sun will be likely to have ejected material in significant quantities, and its mass may reduce to 50 or 60% of its present value. One of the main mass loss stages will occur just before the Sun arrives at position C on figure 4.9. The hot, blue, shrinking Sun will therefore be surrounded by a faintly glowing expanding nebula of thin gas known as a planetary nebula (see Journey 5 and figures 5.1 and 5.2)—a scene that was described at the start of our journey more poetically, though surely coincidentally, by Coleridge almost two centuries ago.

Any remnants of the inner solar system that survived the solar expansion up to the red giant region will also have survived the subsequent changes, and will now orbit the dying Sun. The outer planets may well have survived relatively unscathed, only perhaps losing small amounts of material blown away during the red giant and planetary nebula phases.

The Sun will eventually collapse down and stabilise at a size comparable with the present-day Earth. Though it will have lost some mass, its density will be in the region of 10^9 kg m^{-3}. Such a density can only be achieved if the material is electron-degenerate (see earlier discussion and Journeys 6 and 7). The Sun will have become a white dwarf.

Once the collapse down to a white dwarf finishes, the last available energy source for the Sun will have disappeared. It will therefore slowly radiate away its stored heat, and in a time probably measured in many aeons, cool down to the temperature of interstellar space (a few degrees absolute). Over the same period most of the other, bright, stars in the galaxy will have completed their lives, and the galaxy will be populated by only the dim red dwarfs at the bottom end of the main sequence. Our final view of the solar system therefore has the frozen embers of the planets whirling endlessly around the cold, ultracompressed ashes of the Sun, the whole still faintly illuminated by a fading red glow from the last small stars of the galaxy. As Byron expressed it, again surely coincidently, in his poem *Darkness*:

> *The bright sun was extinguish'd, and the stars*
> *Did wander darkling in the eternal space,*
> *Rayless, and pathless, and the icy earth*
> *Swung blind and blackening in the moonless air.*

Journey 5

A Study in Contrasts

*Truth is like a rabbit in a bramble patch. All you can do is circle
around it, point, and say "It's in there somewhere".*

Pete Seeger

Pick up almost any 'coffee-table' astronomy book, and in glowing reds,
yellows and greens you will find fantastically beautiful photographs
labelled 'planetary nebulae' (figure 5.1). Often they appear roughly
spherical, but many other shapes abound. In truth, they have nothing
whatsoever to do with planets or planetary systems. The misnomer arose
at the beginning of the 19th century. When William Herschel was
compiling his catalogues of nebulae for the Royal Society, he thought
that some of these wispy fragments of gas had a similar appearance to the
planet Uranus. Since then the name has stuck, and it seems unlikely now
that a more appropriate one will be invented.

Substantial though many planetary nebulae seem, they are only just
this side of a complete vacuum. Typically, each cubic metre of a nebula
would contain 10^7 atoms, ions or electrons. This compares with the
'space' around the Earth and in which our low Earth-orbiting spacecraft
move, which has some 10^{10} atoms, ions or electrons per cubic metre.

A space traveller inside a planetary nebula would thus hardly notice
that it was there, apart perhaps from a general glow over the whole
sky which would be fainter than the Milky Way as seen from Earth.
Insubstantial though it is, the material forming the planetary nebula is
still a thousand times denser than the interstellar medium, and heated by
ultraviolet light from an incredibly hot star near its centre, it shines out
like a distant beacon fire in our telescopes.

These rarefied, ghostly objects, however, are the signposts to the
forerunners of some of the densest objects to be found in the universe.
The star near the centre, though sometimes hidden by gas and dust from
our eyes, is likely to be a few tens of times brighter, and rather smaller,

than our Sun. Its colour, if our eyes were sensitive enough to see it, would be an intense blue, and we would estimate its temperature to be 100 000 K or more.

In a 100 000 years or so, the central star will collapse down to the size of the Earth and fade to a small percentage of the Sun's brightness. Its mean density may then be 10^{12} kg m^{-3}: 10^{28} times denser than the nebula! In such a condition, the star will have become a white dwarf (see also the previous Journey).

The nebulae are the most eye-catching objects, particularly on long exposure photographs. They are, however, ephemeral. They will disperse and disappear into the interstellar medium in just 100 000 years or so. The only long-term trace of their passing may be a slight enrichment of the interstellar medium in carbon, nitrogen and oxygen.

The white dwarfs, by contrast, will be around for many aeons, though not all white dwarfs are preceded by planetary nebulae. We shall see them reappearing in the guise of novae, supernovae, binary stars, x-ray sources, polars etc, as well as in the form of isolated faint single stars. Firstly though, we look at the nebulae.

We have already seen some of the more spectacular and better known nebulae (figure 5.1). Many others with a wide variety of shapes and sizes may be found (figure 5.2). These images are obtained using large telescopes and long exposure times; however, they would be quite misleading as a guide for a visual observer who would see little or nothing in most cases, even with a large telescope.

A few of the nebulae are bright enough to be found in small telescopes though. Thus on late summer and autumn evenings the summer triangle, Vega, Deneb and Altair (figure 5.3) is prominent overhead. On a line from Vega to Altair and one fifth of the distance between them may be seen a pair of faint naked-eye stars. Even a 75 mm (3″) telescope pointed midway between these two stars will reveal a faint blur. At first sight it may look like an out-of-focus star, but with continued observation, it resolves into a pale, wispy, greyish smoke ring. It is M57, the Ring Nebula in Lyra (figure 5.1). The nebula is without the colours of the photographs because it is not bright enough to trigger the eye's colour vision, but nonetheless it is incomparably finer seen for oneself than at second-hand.

Such direct images of planetary nebulae can tell us something of their natures. They are probably gaseous, many but not all have a detectable central star, they probably originate from the central star and are still expanding outwards from it in a roughly spherical fashion (the expansion can be seen in a few cases on photographs taken several decades apart), and so on. But confirmation of these data and much more comes only from spectroscopic and radio observations. Indeed, over half of the 1 500 known planetary nebulae actually appear star-like on direct

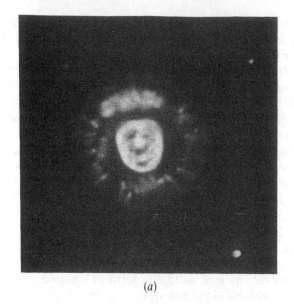

(a)

(b)

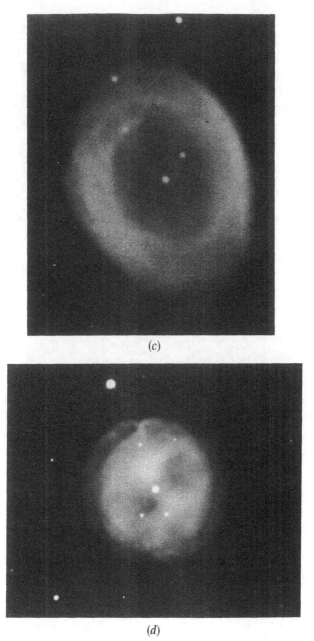

(c)

(d)

Figure 5.1   Some well known planetary nebulae. (a) Eskimo (NGC 2932), (b) Helix (NGC 7293), (c) Ring (NGC 6720, M57) and (d) Owl (NGC 3587, M97). ((a)–(c) reproduced by permission of Hale Observatories, and (d) by permission of Mount Wilson and Palomar Observatories.)

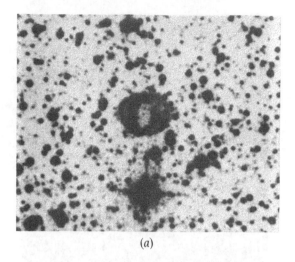

(a)

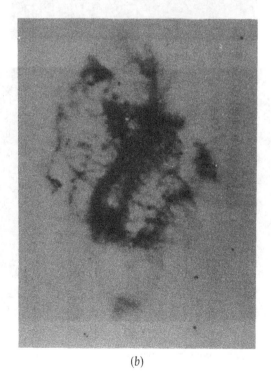

(b)

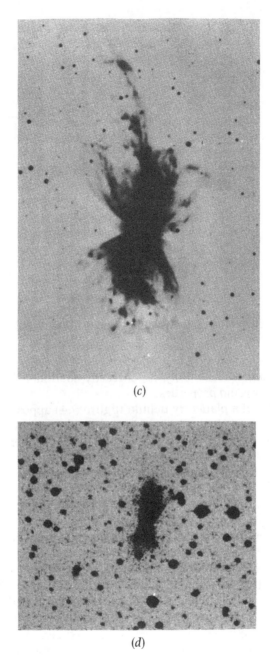

(c)

(d)

Figure 5.2 Some less well known planetary nebulae (negative images). (a) Zodet's planetary, (b) NGC 5189, (c) NGC 6302 and (d) ESO 172-?07. (Reproduced by permission of the European Southern Observatory.)

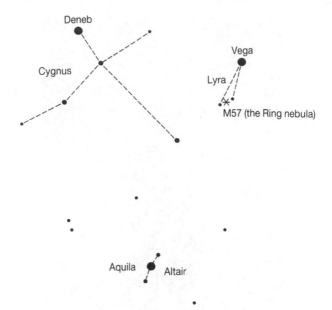

Figure 5.3   The summer triangle and the Ring nebula (M57).

photographs, and so can only be correctly identified by their optical spectra and their radio properties.

The spectrum of a planetary nebula (figure 5.4) appears most odd to a stellar astrophysicist. Instead of the usual bright continuum with dark absorption lines, bright emission lines predominate. There is little or no continuum, and so no absorption lines. The nebula's spectrum is rather similar at first sight to the spectrum of an emission lamp such as the ubiquitous sodium street lights. This, of course, is because both spectra are produced by thin, hot gases. At second sight, however, the nebula spectrum reveals itself as very strange: many of the emission lines, including the brightest ones, do not coincide with any of those commonly associated with the known elements.

This latter oddity became a major problem in the years after spectroscopes were first turned onto planetary nebulae in the late 19th century. At one stage, a new element, nebulium, was even invented to try and account for the lines. Eventually, the true explanation was found. The lines *are* due to known elements, particularly oxygen, neon and sulphur. The probability of their occurrence under 'normal' (i.e. laboratory) conditions is so low however that they are known as forbidden lines. Only in the very rarefied, low energy environment of the nebulae do they become significant, and in many cases they then dominate the resulting spectrum.

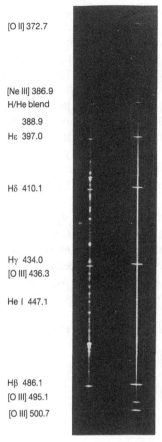

[O II] 372.7

[Ne III] 386.9
H/He blend
388.9
Hε 397.0

Hδ 410.1

Hγ 434.0
[O III] 436.3

He I 447.1

Hβ 486.1
[O III] 495.1
[O III] 500.7

Figure 5.4 Spectra of planetary nebulae: (a) IC 418 and (b) BD + 30° 3639. (Reproduced by permission of the Royal Astronomical Society.)

Apart from atomic parameters, the strengths of the spectrum lines, whether normal or forbidden, are dependent mostly on the temperature and density of the nebula. This conversely allows the astrophysicist to use the relative line strengths to estimate the values of these two quantities.

Thus it is that we find the typical density for the nebulae to be 10^7 atoms, ions or electrons per cubic metre. We also find the temperature to be in the region of 10 000 K, nearly twice as hot as the surface of the Sun. However, we must stop and ask what we mean by this latter figure. If our earlier hypothetical space traveller were to put a hand (or tentacle) out into this very hot gas, then it would drop off from frost-bite rather than through vaporisation. The explanation of this paradox lies in the low density of the gas. The temperature of 10 000 K which we have measured is the kinetic temperature, i.e. that corresponding to the random motions of the particles. These are about 15 km s^{-1}. But there are so few particles, that the effective temperature (loosely, the temperature that you would feel upon sticking a hand outside the spacecraft) is only a few degrees above absolute zero.

The radio emission from the nebulae arises from a process known as free–free or bremsstrahlung radiation. Such radiation occurs when an electron, set free from an atom by ionisation, passes close to another nucleus. During the passage it may slow down with respect to the nucleus, though without being captured by it. The reduction in the electron's kinetic energy reappears in the form of a free–free photon (figure 5.5).

The radio spectrum of the nebula typically shows a rise in intensity with frequency, which eventually flattens out (figure 5.6). The flat part of the curve occurs for those frequencies at which the nebula is partially transparent, while over the frequencies covered by the sloping portion of the curve, the nebula is opaque. The frequency of the shoulder at which the nebula becomes opaque provides another means of estimating the nebula's temperature.

We have said that the nebulae are expanding, and indeed it is obvious

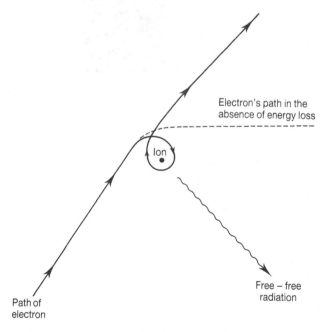

Figure 5.5   The production of free–free radiation.

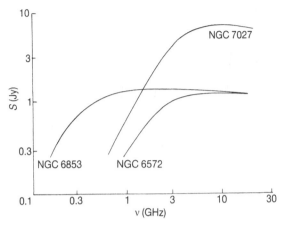

Figure 5.6   Radio spectra of planetary nebulae. (Reproduced from IAU Symposium no 34 (1968) p. 87, Y Terzian, by permission.)

that this must be the case if the nebula has originated from the central star. The apparent proof of this seems to be shown on direct images obtained several decades apart when some nebulae may be seen to have increased in size. Thus NGC 6572 expanded by about 2.5″ in the 45 years between 1916 and 1961, which would correspond to a linear velocity of about $100\,\mathrm{km\,s^{-1}}$. This would seem to be clear-cut evidence of the nebula's physical expansion. However, it is always possible that the gas forming the nebula is actually stationary, and it is just that cool gas outside the visible nebula is being heated to visibility. Definite proof of the expansion therefore only comes from spectroscopy. The Doppler shifts of and within the spectrum lines lead to an estimate of the average expansion velocity for the nebulae of $20\,\mathrm{km\,s^{-1}}$. The average planetary nebula with a diameter of one light year is thus about 10 000 years old.

It is generally accepted that the nebula is formed from material expelled from the central star, but the precise mechanism whereby this happens is still not completely clear. The theory of the formation of planetary nebulae most widely accepted at present suggests that is the cool red giants known as long period variables, such as Mira (figure 5.7), which generate them. The material forming the nebula may be the surface layers of the star pushed outwards by the pressure of radiation from the star. It seems more likely that the surface layers of the star become unstable as their total energy exceeds their potential energy. Thus as the material expands (for whatever reason—perhaps just random turbulence) the gas cools, releasing ionisation energy which then powers further expansion, and so on. In either case, the central star loses its outer layers to form the nebula.

One of the reasons for the uncertainty about the early stages of the

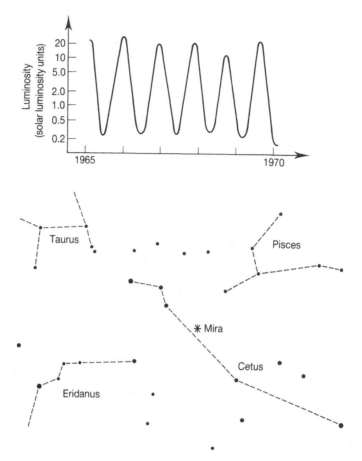

Figure 5.7   o Cet (Mira): light curve and finder chart.

formation of planetary nebulae is that they are hidden from us by dust condensing out around the cooling star. Recently, however, infrared observations have started to reveal some candidates for this stage of the process. The amount of material lost from the star is estimated to range from a fifth of a solar mass for a star with a mass similar to that of the Sun, to two solar masses when the original star has a mass three times that of the Sun. In all cases therefore, the mass of the central star is expected to be brought below the upper limit of 1.4 solar masses (the Chandrasekhar limit, Journey 7) for the mass of a white dwarf.

The complex structure of some of the nebulae (figures 5.1 and 5.2) is probably due to a variety of processes. The central star may be a member of a binary or multiple system, and the orbital motion may then modulate the mass loss, perhaps to produce a nebula such as the Helix.

Alternatively, the star may have a rapid rotation or strong magnetic field which favours mass loss in some directions over others, producing bipolar and other non-spherical nebulae. Then the local interstellar medium may be inhomogeneous or contain small gas clouds which interfere with the expansion. Finally, in many cases the central star seems to develop a high velocity stellar wind ($1\ 000\ \mathrm{km\,s^{-1}}$) after it has lost most of its material. This sweeps up and compresses the inner parts of the nebula leading to ring (actually spherical) nebulae such as M57 in Lyra (figure 5.1).

Let us now journey on towards the central stars themselves. They are generally physically small: half the size of the Sun or less on average. Their temperatures are very high, however: 30 000 to 150 000 K or more. The central stars' luminosities are therefore typically 50 or so times that of the Sun.

The high temperature of the central star is the reason why we are able to see the surrounding nebula. At such high temperatures, much of the star's emission is in the ultraviolet region of the spectrum. Radiation with wavelengths shorter than 91.2 nm has enough energy to ionise hydrogen atoms and so will be absorbed by the material of the nebula, since this is mostly hydrogen. When the free electrons which result from these ionisations recombine with ions, they will generally do so to an excited level. The electrons then cascade down to the ground state, emitting radiation into some of the atom's spectrum lines. Thus it is that we get the emission line spectrum of the nebula.

The central stars are not at this stage white dwarfs. They seem in fact to divide into four types on the basis of their spectra. The first of these are called the Wolf–Rayet type. Now most Wolf–Rayet stars are far more massive than the stars at the centres of planetary nebulae. Central stars of this type therefore just mimic the features of true Wolf–Rayet stars without actually being in that category. The distinguishing features of Wolf–Rayet stars and of central stars of the same type is that their spectra comprise a high temperature continuum which has very broad and very strong emission lines superimposed. The lines are largely due to helium and nitrogen (the WN subclass) or helium, carbon and oxygen (the WC subclass). In neither subclass is there much evidence of the presence of hydrogen. Additionally, the shapes of the spectrum lines, especially in the ultraviolet (figure 5.8) suggest that the stars are losing mass at rates of up to 0.01% of a solar mass per year in the form of a very high velocity ($2\ 000\ \mathrm{km\,s^{-1}}$) stellar wind.

The logical, though not necessarily correct, explanation for Wolf–Rayet-type central stars would seem to be that they are just the cores of the original stars which have been revealed by the loss of the stars' outer layers to form the nebula. The composition of the central star would then just reflect the products of the last significant nuclear reactions, the

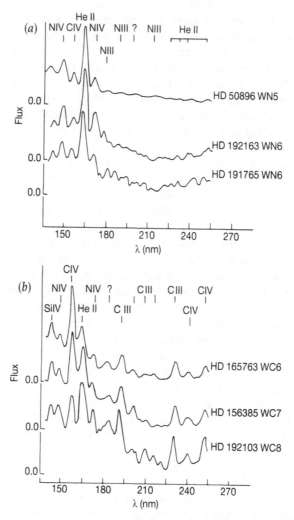

Figure 5.8   Ultraviolet spectra of (*a*) WN and (*b*) WC stars. (Reproduced from *Mon. Not. R. Astron. Soc.* **182** 559 (1978) A J Willis and R Wilson by permission.)

carbon cycle in the WN stars, and helium burning (the triple-alpha reaction, Journey 4) in the WC Stars.

A second type of central star has a spectrum very similar to that of the normal O-type stars, though again, its mass must be very much less than the 20 to 40 solar masses usual for such stars. A third group is labelled Of-type and is intermediate in its properties between the Wolf–Rayet and the O-type central stars. The final group is difficult to study because its optical spectrum contains no observable spectrum lines. It is therefore called the continuum type of central star.

Quite rapidly, in 100 000 years or so, the nebula will disperse, mix with the interstellar medium and disappear from view. Over a similar timescale, the central star will shrink in size while remaining roughly constant in temperature. Its optical luminosity will fall during this time, but there is likely to be a rise in the energy emitted in the form of neutrinos (Journeys 4 and 7) until this dominates the energy loss from the star.

Eventually, at a radius in the region of 10 000 km and at an optical luminosity about 0.1% that of the present Sun, the central star will settle down to its old age as a white dwarf. Thereafter it will continue to diminish in brightness, but this time because the star's temperature will fall as it radiates away its stored heat, there being no significant nuclear energy generating reactions left inside the star by this stage (see also Journey 4). This final fading is a slow process, and it seems likely to take in the region of 10^{10} years for each further reduction in luminosity by a factor of 10.

Not all white dwarfs have planetary nebula precursors. Perhaps only one in four will have gone through that route; the others are likely to be members of close binary systems. Then the mass lost prior to the white dwarf forming is likely to have accreted onto the companion star rather than going to form a nebula. These close binary white dwarfs often go on to evolve in spectacular ways, but in order to understand why that occurs we first need to look in a little more detail at the structure of white dwarfs and at the properties of the electron-degenerate matter of which they are largely composed.

We have seen that a white dwarf may be only the size of the Earth, and yet still have a mass similar to that of the Sun (Journey 4). This would give it an average density of 4×10^9 kg m^{-3}: about four million times the density of water. Clearly such material must be very different from that which we are accustomed to on Earth. Indeed it forms what is often called the fifth state of matter. We are more or less familiar with the first four: plasma (ionised gas), gas, liquid and solid. The next stage in this sequence is known as electron-degenerate matter. The sixth and last stage before black holes is baryon-degenerate matter. It is the stuff of neutron stars, and it may reach densities of 10^{17} kg m^{-3}, but it will not further concern us now (see Journey 7).

Electron-degenerate matter (see also Journeys 4, 6 and 7) is peculiar in many ways. Most significantly, it differs from ordinary matter in that its internal pressure is independent of its temperature. If this sounds unexciting, think of what life would be like if ordinary matter behaved in the same way: there would be no petrol, diesel or steam engines, fires would smother themselves, saucepans on the stove would burn their bottoms out, hot air balloons would not fly, there would be no wind or rain, and so on almost *ad infinitum*.

The first result of this lack of temperature dependence for the pressure in degenerate matter is that there is an upper limit (the Chandrasekhar limit to which reference has already been made) for the mass of a white dwarf. This is because the pressure (strictly only the electron pressure, but this usually dominates) depends only upon the density of the material. The pressure is proportional in fact to the 4/3 power of the density. Since the pressure inside a more massive star must be greater than that inside a lighter star, in order to balance the increased weight of material we find that the more mass a white dwarf has, the smaller it must be. Thus at a quarter of a solar mass a white dwarf would be about 20 000 km across, falling to 8 000 km at one solar mass and so on. At 1.4 solar masses, the diameter would (theoretically) have shrunk to zero, giving this as the maximum possible mass for a white dwarf.

The earlier mean density calculation hides a wide range of actual densities in white dwarfs (figure 5.9). The central density may be 10 to 20 times the average density, while near the surface almost-normal matter may be encountered. There is even likely to be a thin atmosphere of ionised gases.

This, then, is the likely end for the central stars of planetary nebulae. Since they will probably be single stars or at most members of very widely separated binary systems, they will just settle down to cool slowly (like the Sun, Journey 4). In some tens of thousands of millions of years, or longer than the present age of the universe, they will reach the

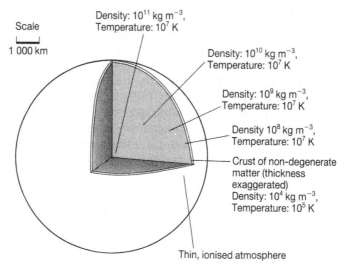

Figure 5.9   Structure of a white dwarf of one solar mass.

temperature of interstellar space. During this cooling stage they must spend some aeons near 'room' temperature. We thus have the intriguing speculation of the possibility of carbon-based life evolving on their surfaces during that time. With mountains equivalent to Everest being only 20 mm high, meteorites landing with kiloton explosive forces and the life-forms themselves probably having to resemble flimsy typing paper, it would certainly be an exotic environment.

Returning from science fiction to science fact, we now look at how being in a close binary system affects the later stages of a white dwarf's life. In 1844 Bessel predicted that Sirius was a binary system because the visible star had a sinuous motion across the sky (figure 5.10). But its

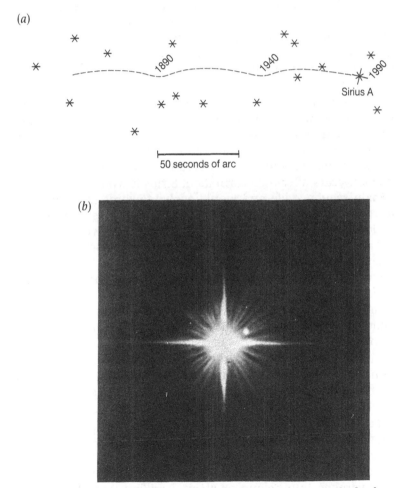

Figure 5.10   Sirius. (a) The movement of Sirius A against the background stars. The 'wobble' is caused by the gravitational pull of Sirius B. (b) Sirius A and B. (Reproduced by permission of Prof. W. J. Kaufmann.)

companion was not found for another 18 years. The reason for this long delay became apparent when Alvan Clark (a 19th century American telescope maker) saw the companion for the first time. It was nearly 10 000 times fainter than the primary star, and hence extraordinarily difficult to see because of the glare from the primary.

We now know that Sirius A is a main sequence star with a mass 2.14 times that of the Sun, while Sirius B is a white dwarf with a mass 1.05 times that of the Sun. These figures immediately suggest a paradox. Theories of stellar evolution give a maximum age for Sirius A of 10^9 years, while Sirius B has a minimum age of 10^{10} years. If Sirius were the only known example of such a situation, then we might accept that the two stars had originated separately and at different times, and had come together later to form the binary. Such a mutual capture, however, is a very low probability event requiring a third body to be involved to remove the excess energy. We would only expect a few (0 to 10) such events to have occurred throughout the whole volume and lifetime of the Milky Way galaxy.

But Sirius is only one of thousands of similar and related systems, so that some other explanation must be sought. The true answer is thought to be that the system did form originally as a binary, and that Sirius B was originally the more massive star. Sirius B would then have been about three solar masses and Sirius A perhaps one solar mass. A few hundred million years after their formation, Sirius B would come to the end of its main sequence life and would evolve towards the giant and supergiant regions. As its size increased, it would start to encroach into the gravitational sphere of influence of Sirius A. Material would then be lost from Sirius B, most of which would accrete onto the surface of Sirius A, but some might disappear from the system altogether. Eventually so much mass would have been lost from Sirius B that its nuclear energy generating reactions would cease, and it would start its collapse down to a white dwarf. Sirius A meanwhile would have doubled its mass to its present value and be the brighter and more massive star in the system, just as we now see it.

At some future date, of course, Sirius A must come to the end of its main sequence life and in its turn evolve towards the giant region; mass will then be transferred back to Sirius B. The cycle does not simply repeat itself, however. The results of this second phase of mass exchange are quite varied and can be very spectacular.

When matter falls from a distance onto the surface of a white dwarf, it releases a vast amount of energy: about $3 \times 10^{13} \, \mathrm{J \, kg^{-1}}$.

That energy is about 5% of the energy released in converting one kilogram of hydrogen into helium or about a million times the energy released in typical chemical reactions. Unlike the nuclear reactions though, this energy is released at the surface of the star where it may

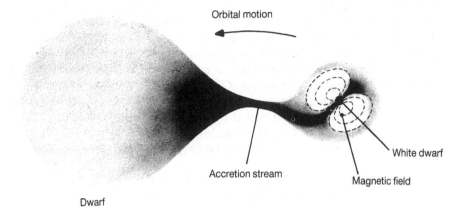

Orbital motion

White dwarf

Accretion stream

Magnetic field

Dwarf

Figure 5.11   Schematic model of an AM Her type system.

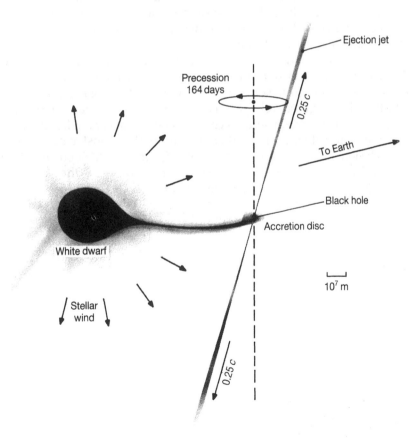

Ejection jet

Precession
164 days

0.25 c

To Earth

Black hole

Accretion disc

White dwarf

10^7 m

Stellar
wind

0.25 c

Figure 5.12   A model of the SS433 system.

produce immediately observable effects. The type of object which we may then see depends upon the details of the mass exchange, the composition of the white dwarf, and the presence or absence of magnetic fields etc.

Thus we have the AM Her type stars, which are also known as polars from the strong and variable polarised component to their emission, and which are also weak x-ray sources. These seem to be dwarf star and white dwarf binary systems. Material from the dwarf is accreting onto the white dwarf where it is channelled by the latter's strong magnetic field. The various optical and x-ray emissions are then largely due to the impact of the accretion streams onto the white dwarf's surface (figure 5.11).

Somewhat similar systems to the AM Hers stars probably result in dwarf novae and classical novae, as we shall see on our next journey. We also have the fascinating system SS433 (figure 5.12), where a white dwarf is being pulled apart by the gravitational field of a neutron star or black hole. The impact of the accretion streams from the white dwarf with the other object in the system results in a small proportion of the material being flung out into two jets in opposite directions and at velocities of up to $75\,000\ \mathrm{km\,s^{-1}}$.

Finally, accretion in a binary system may take the mass of the white dwarf over the 1.4 solar mass Chandrasekhar limit. It must then collapse down to a neutron star, or perhaps even to a black hole. The gravitational energy so released, combined with nuclear energy from the fusion of the carbon and oxygen which probably make up the bulk of these white dwarfs, may then result in that most cataclysmic stellar explosion of all, a type I supernova (Journey 7). Such an object may reach a thousand million times the brightness of the Sun for a few months: a single star shining as brightly as an entire galaxy. Not a bad effort from what is sometimes one of the least regarded of stars!

Journey 6

Thar She Blows!

The heavens themselves blaze forth the deaths of princes

Julius Caesar
W Shakespeare

We have travelled far enough and wide enough through time and space to realise by now that the 'eternal unchanging fixed stars' are neither eternal, nor unchanging, nor fixed. Nonetheless over a human life span most of the changes are going to be miniscule. Even the ephemeral planetary nebulae of the last journey will generally exist for a hundred thousand years or more. However, not all aspects of stellar evolution take place at a snail's pace. Some dramatic changes can occur over intervals ranging from milliseconds to a few days, as we shall see on this and on the next of our journeys.

Over the two or three millenia prior to 1610, when astronomical observations relied only on the unaided eye, the heavens generally did appear eternal and largely unchanging. There were some changes though, to be seen. The planets for example moved slowly with respect to the background stars. Another change was found as early as 130 BC, when Hipparchus discovered that the point in the sky around which the heavens appear to rotate (the celestial north pole, currently close to the Pole Star) was moving slowly through the stars. In fact it traces a huge circle through the northern constellations (figure 6.1) taking about 25 750 years to complete a circuit. Only for a small fraction of that time therefore do we have a bright star (Polaris) close to the pole, and only two other prominent stars, Vega and Deneb, come anywhere near its path. Neither of these changes, however, are to the stars themselves.

Occasionally and randomly though, new objects would appear in the sky, usually to disappear again a few weeks or months later. The most visually dramatic of such events, the comets (figure 6.2) would change their shape and brightness and move against the background stars.

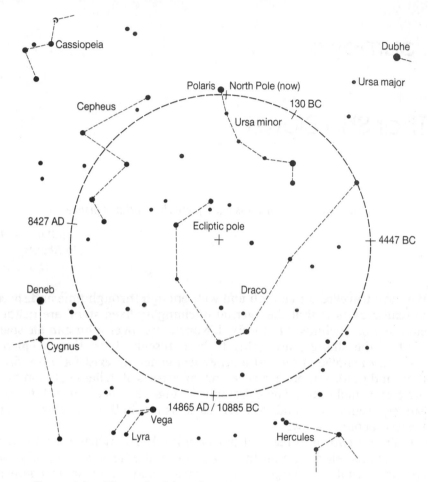

Figure 6.1   The path of the celestial north pole through the sky due to precession.

However, they were generally thought to be phenomena within our atmosphere. Not until the bright comet of 1577 did measurements by Tycho Brahe prove them to lie at least beyond the Moon. Such apparitions therefore, though they caused consternation and dread and were taken as dire warnings, were usually thought of as a part of the corrupt Earthly sphere, not a part of the immaculate heavens. Only a new star, Stella Nova, gave indubitable evidence of change beyond the solar system. Such new stars could appear a few times a century, brighten and then fade into invisibility again a few months later. We know them now, from that earlier term, as novae.

Just as the behaviour of novae differs from the normal run of stellar

Figure 6.2   Comet West, 1976. (Reproduced by permission of the Royal Observatory, Edinburgh.)

processes, so the astronomers who look for them form a different breed from the rest. The ambition of most observational astronomers (whether professional or amateur) is to use larger and larger telescopes to peer ever deeper into the heavens. But the hunter of novae requires to be armed only with a large pair of binoculars, a deck chair, and an exhaustive knowledge of the sky. He or she probably has more in common, in terms of attitude at least, with the mast head look-out of an old sailing whaler than with most modern astrophysicists.

The nova hunter, who is usually equally happy to discover comets, just scans the sky on every clear night, lying comfortably in the deck chair, searching for the star or the faint fuzz of an incoming comet that was not present the last time he or she scanned that part of the heavens. It is one of the few areas left in astronomy where the amateur can make a real contribution. Of course, once the nova (or comet) has been discovered, the professional astronomers with their large telescopes muscle-in. But the discoverer still has had the thrill of discovery, and in the case of comets, the reward of having it named after him or herself.

If the few novae per century that become bright enough to be seen with the naked eye were all that were available, then the patience of even nova hunters would be sorely tried. Fortunately there are about 30 to 40 novae occurring in our galaxy each year. About 10% of these novae will become visible from the Earth, though most will remain detectable only via telescopes even when at their brightest. Telescopes have also shown that they are not really 'new stars' appearing where no star existed previously, but faint stars that have become very very much brighter (figure 6.3).

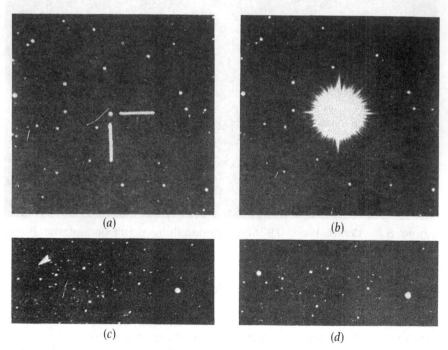

(a)   (b)

(c)   (d)

Figure 6.3   (a) Nova Herculis 1934. (Reproduced by permission of the Lick Observatory.) (b) Nova Cygni 1975. (Copyright Ben Mayer, Problicom Sky Survey, from *Contemporary Astronomy* by J M Pasachoff.)

A typical nova then, before its explosion, would appear a few times brighter than the present Sun. At a fairly typical distance of 10 000 ly, it would just be visible through a 0.5 m (20″) telescope. Careful observation might show that it was actually a close binary: two stars orbiting their common centre of mass. The orbital period of the binary would typically be less than a day. Even more careful observation might show that the system was 'flickering': changing its brightness by 1% to 10% every minute or so.

The prenova's spectrum, if we had a large enough telescope to be able to study it, would show the system to be very hot: 20 000 to 50 000 K, and there might be faint emission lines coming from thin hot gas in addition to the more normal absorption lines. Abruptly and unpredictably the prenova system would start to brighten rapidly, beginning the nova outburst. In just one or two days it would be brighter by some 10 000 times. Often it would then pause for a few days before brightening again by another factor of three or four. At its maximum a nova may be 50 000 times brighter than the Sun, with the occasional exceptional example reaching a million solar luminosities.

If the Sun were to become a nova (which as we shall see shortly it cannot), then all the planets out to at least Uranus would be vaporised. Perhaps fortunately, it seems unlikely that stars which do become novae can also possess planetary systems.

At its maximum brightness, a typical nova, which might be 10 000 ly away, would be just approaching visibility to the naked eye. Most novae, however, still require a small telescope or pair of binoculars in order to be found. It is thus only the rare, exceptionally bright or nearby nova that becomes easily visible.

After a few days at maximum brightness, the nova will start to fade. Its changes occur much more slowly now, declining in brightness by a factor of two over a period ranging from a week to two or three months (figure 6.4). Some, but not all, novae have large fluctuations in brightness lasting for months, superimposed on this general decline. After a few years, the nova will have returned to its pre-outburst brightness. Closer examination then will show it to be substantially the same as before the outburst in its other properties as well.

The spectrum also alters radically throughout the nova outburst. Perhaps surprisingly, given the apparent explosive nature of the nova, the first change is towards lower temperatures. Then the emission lines strengthen and soon come to dominate the spectrum. Indeed, soon after maximum brightness the spectrum may consist only of the emission lines. All the lines, emission as well as absorption, are found at shorter wavelengths than normal, indicating that the material producing them is moving towards us at between 500 and 2 000 km s^{-1}.

To begin with, the effects of the nova's outburst are primarily seen in

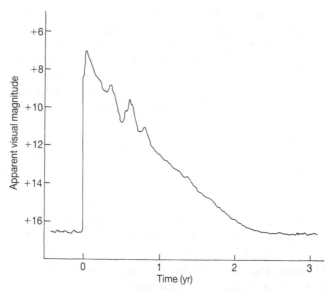

Figure 6.4   Schematic light curve of a classical nova.

the visible region of the spectrum. However, as the visible radiation starts
to decline, first the ultraviolet and then the infrared emissions strengthen,
coming to dominate the energy emission after a few months. The total
amount of energy emitted over the few years of the outburst may reach
10^{39} J: enough energy to power the Sun for 100 000 years. Such an
event would normally be called a classical nova. Two other types of nova
are also recognised: the recurrent novae and the dwarf novae.

The recurrent novae are easily described, for in most ways they
resemble a classical nova, but two or more outbursts have been seen
from them. The separation of the outbursts ranges from ten years to a
century. Their other difference from classical novae lies in the appearance of their spectra between outbursts, which are like that of a fairly
normal red giant.

The dwarf novae are rather different phenomena. Their outbursts are
much smaller, increasing in brightness by perhaps 'only' a factor of a few
hundred. They then fade again quickly, so that the outburst lasts for just
a few days or weeks. A month or two later, though, the whole process is
repeated, and so a light curve like that shown in figure 6.5 is obtained.
This is so different from the behaviour of the other novae that it is clearly
a different phenomenon. Nonetheless, there are many similarities
between dwarf and other novae outside one of their outbursts. They are
thought to be such basically similar systems that a classical nova
outburst may be expected to occur on a dwarf nova at some stage in
addition to its repeated minor outbursts.

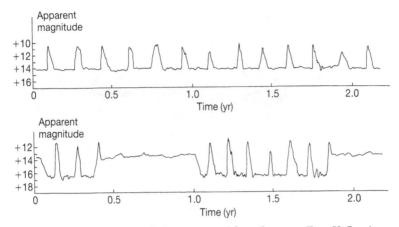

Figure 6.5   Schematic light curves of dwarf novae. Top: U Gemino-
rum type stars. Bottom: Z Camelopardis type stars.

Let us turn then to seeking an explanation for the nova phenomenon.
Like the properties of white dwarfs, the pattern of the evolution of the
Sun which we encountered previously and the supernovae to be
encountered on the next journey (Journeys 4, 5 and 7), the events
leading to classical and recurrent novae arise from the properties of
electron-degenerate matter. The important property again is that the
pressure inside a degenerate material is (largely) independent of its
temperature.

Imagine a close binary system in which one of the stars is a white
dwarf. A system, for example, such as Sirius A and B (Journey 5). The
stars, however, are much closer together than in Sirius, so that both the
white dwarf, its companion and their orbits could fit inside the Sun. The
companion star to the white dwarf at some stage must come to the end of
its main sequence life. Then just like the Sun (Journey 4), it will expand
as it evolves towards the giant star region of the Hertzsprung–Russell
diagram. Unlike the Sun though, the companion will not be free to continue the expansion indefinitely, for its surface layers will eventually
encroach into the gravitational sphere of influence of the white dwarf.
Material would then be lost from the companion, spiralling down
towards the white dwarf and accumulating in an orbiting disc, the
accretion disc, surrounding it (figure 6.6).

Large amounts of potential energy will be released during this process:
some 3×10^{13} J for every kilogram that falls onto the surface of the white
dwarf. This is 5% of the energy produced in the nuclear reactions
converting one kilogram of hydrogen into helium. Unlike the nuclear
reactions though, this energy is produced outside the star where it may
have immediately observable effects. In fact, most of it is released at the

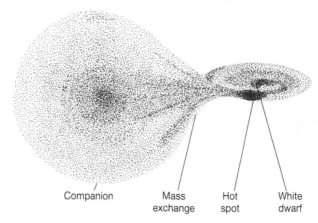

Companion   Mass   Hot   White
             exchange   spot   dwarf

Figure 6.6   Mass exchange in a white-dwarf-containing close binary system.

point where the material from the companion (the accretion stream) collides with the accretion disc. At the impact point a hot spot is produced, which is brighter than either of the stars. Thus it is this hot spot that is observed when the nova is quiescent.

The material within the accretion disc, which is mostly hydrogen from the surface layers of the companion, will be in a very turbulent state, and so it will slowly spiral down to accumulate on the surface of the white dwarf. As time passes, this surface layer of hydrogen will become thick enough for its lower parts to become electron-degenerate. Eventually the temperature and pressure within the surface layer will become high enough to support hydrogen fusion (Journey 3). When this happens the temperature will rise very rapidly, but as we have seen the pressure will remain unchanged since it is independent of temperature in electron-degenerate material. The surface layer will therefore remain undisturbed.

Now the rates of hydrogen fusion reactions are very temperature-dependent. Doubling the temperature may cause the energy output to rise by a factor of 100. The fusion reactions therefore start to runaway, and the temperature in the outer layer may shoot up to 100 000 000 K, releasing some 10^{38} J. Eventually the pressure from the non-degenerate nuclei, which does still depend upon the temperature, will disrupt the outer layers. The material will then revert from being electron-degenerate back to 'normal'. The electron pressure will regain its temperature dependence, and the resulting pressure pulse will convert the disruption into an explosion which rids the white dwarf of its blanket of hydrogen. This explosion we may then observe and call a nova.

The actual increase in brightness of the system arises from the increased area of the surface that is emitting radiation as the shell expands outwards. The hot spot, which had previously been the main radiation source, is disrupted by the explosion, and so the spectrum changes towards the lower temperature which characterises the outermost layers of the shell. Later, as the shell expands, its spectrum comes more to resemble that of a hot nebula like the planetary nebulae and HII regions visited earlier. This expanding shell can sometimes be seen directly many years later as it reaches a size to be resolved from Earth (figure 6.7).

The later decline in the optical brightness occurs as the surface emission decreases, with the wilder fluctuations (figure 6.4) arising from changes in the transparency of the outer layers.

For all the spectacular violence of the nova explosion, it does not change the basic structure of the original close binary system. Before the shell has cleared sufficiently to allow the stars to be seen directly, within a year or so of the outburst, the expansion of the companion will resume. The accretion stream will reform and in due course produce a new accretion disc. By the time we can see through the expanding shell, we once again see the hot spot where the accretion stream and disc collide, rather than either of the stars.

The material in the new accretion disc will again accumulate onto the surface of the white dwarf through losing energy via viscous interactions, and a new layer of hydrogen will start to build up. After perhaps 10 000 years, perhaps 100 000 years, there will be another nova explosion.

Several hundred such explosions might occur within a single system before the companion finally loses sufficient mass to halt its expansion and bring the process to an end. The two stars will then settle down as a pair of white dwarfs in a close binary system. As a final twist, according to some theories, they could then end up as an even more spectacular stellar explosion: a type I supernova, but that story is reserved for the next journey.

To complete this journey, we still have the recurrent and dwarf novae to explain. The recurrent novae are just those novae with periods short enough for two or more outbursts to have been recorded in recent centuries, for all novae, as we now see, are recurrent. The reasons for their much shorter periods are not certain. It may be that they form in more widely separated binaries. The companion can therefore evolve further away from the main sequence before losing mass to the white dwarf. Then it may lose mass more quickly, and the material it loses may be enriched in the isotope helium-3, which would enable the thermonuclear reactions to start at lower temperatures and pressures than is possible for hydrogen and helium-4. Both these effects would tend to lead to more frequent, possibly smaller, outbursts. Another difference from the classical novae is that the recurrent novae are embedded in dense

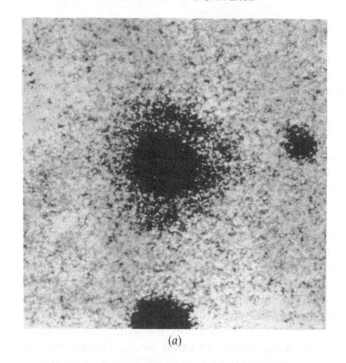

(a)

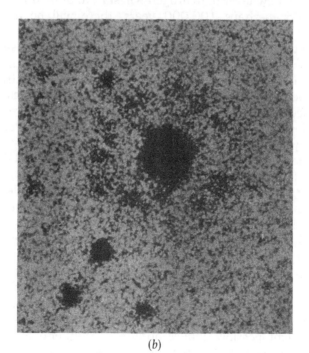

(b)

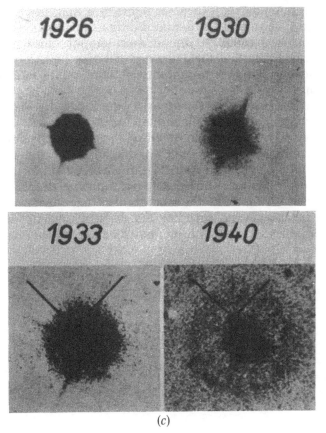

(c)

Figure 6.7 Expanding shells of gas around old novae (negative images). (a) T Pyx (1966), (b) V603 Aql (1918) and (c) CP Pup (1942). (Reproduced by permission of the European Southern Observatory.)

envelopes of gas, though how this affects their outbursts is not clear.

Dwarf novae are more of a problem. They seem to be the same basic type of system as the other novae: a close binary with a companion losing mass to the white dwarf. The energy that powers their outbursts probably comes from the potential energy released by the infalling accretion stream, rather than from thermonuclear runaway explosions. Somehow, perhaps by storage in magnetic fields, perhaps by a positive feedback system wherein increased radiation from the hot spot increases the rate of mass loss from the companion, so increasing the temperature of the hot spot etc, this energy is released in bursts rather than continuously. Some slight support for the model comes from the sub-type of the dwarf novae, known as the Z Camelopardalis stars (figure 6.5). In these stars the outbursts sometimes come to a halt, leaving the star's output almost constant. It is then found that the output during such a constant phase is equal to the average output during a series of outbursts, suggesting that the difference between the two phases of the star's behaviour lies just in an interruption of the energy storage mechanism.

Whatever the cause of the dwarf nova's outbursts, the energy released is insufficient for much of the material in the accretion disc to be expelled from the system, as happens in the other nova types. The material must therefore continue to accumulate on the surface of the white dwarf whilst the outbursts are occurring. Eventually therefore, we may expect a 'normal' nova explosion, a thermonuclear runaway, thousands of times more energetic than the dwarf nova outbursts, to occur. As candidates to watch for the likely next occurrence of a classical nova, the dwarf novae are probably therefore the favourites on which to put your money.

Journey 7

The Hottest Spots in the Universe

Then is it reasonable to think that one can see, by looking in a
microscope, what is going on in another planet?
The Father
J A Strindberg

How much less reasonable would have seemed the thought that peering
into huge tanks of water hidden deep under the Earth's surface would
reveal the inner workings of vast stellar explosions occurring tens of
thousands of light years away.

Yet just such a remarkable observation took place soon after 7.30 am
GMT on the 23rd February 1987. Furthermore it was made by nuclear
physicists, rather than by astronomers. Two groups of workers, one
Japanese, the other American, whilst actually trying to detect the decay
of protons into positrons, detected a total of 19 neutrinos and anti-
neutrinos over a period of just 13 seconds. The direction of some of these
particles could be measured, and they pointed back in space towards the
nearest of our galactic neighbours, the Large Magellanic Cloud.

Neutrinos, as we saw in our first and third journeys, are particles
which differ from nothing at all by only the smallest of margins. They
are continually flooding our whole environment in their thousands of
millions of millions due to the nuclear reactions at the centre of the Sun.

Why then, were these 19 particles of such significance? The answer
became apparent some 18 hours later when a bright new star, a supernova, was discovered in this same Large Magellanic Cloud (figure 7.1)†.

As we saw on our last journey some stars can brighten briefly into
naked-eye visibility and change the patterns of the constellations.

---

† In strict order of precedence, the supernova was discovered optically first. Only
after this did the nuclear physicists look back through their as-then unexamined
data and find the neutrino pulse.

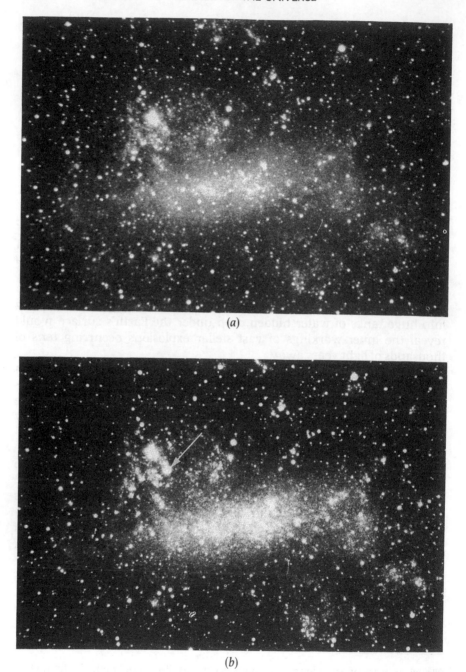

(a)

(b)

Figure 7.1   Supernova SN1987A (arrowed) in the Large Magellanic Cloud; an anomalous type II supernova. (Reproduced by permission of the European Southern Observatory.)

Occasionally one such might be bright enough to outshine all heavenly bodies save the Sun and Moon. Then its importance to the ancients was so momentous that we may still find records of it.

Thus a new star bright enough to be seen in daylight was noted by Chinese astronomers in 1054 in the area of the sky we now call the constellation Taurus. An even brighter object was recorded in 827 AD, somewhere near Scorpius by Arabian observers. Similarly, some South American rock carvings may point to the occurrence of a new star in Vela at a time before the invention of writing. Then, of course, there are the two well attested events witnessed by Tycho Brahe and Johannes Kepler in 1572 and 1604 within the constellations of Cassiopeia and Ophiuchus respectively.

Such new stars could be a nearby example of one of the various types of nova which we visited on the last journey. In fact, despite the superficial similarity between the light curves, they are quite different types of object which we now know as *supernovae*.

The supernovae divide into two main types on the basis of their spectra and light curves, labelled types I and II. We are going to devote most of this journey to following through the events leading to type II supernovae, but we have time for a small diversion to look at the type I supernovae as well.

Type I supernovae are the more spectacular of the two types, reaching a peak brightness 2×10^9 times that of the Sun, or comparable with the entire emission from a small galaxy (figure 7.2). The excursions are by at least a factor of 10^8, and the rise in brightness occurs at a rate of 20% to 50% per day. After a week or so at maximum brightness, the decline occurs at an initial rate of halving in brightness each week, but this soon slows to halving every ten weeks or so. The spectra of type I supernovae are characterised by the absence of lines due to hydrogen, and by line shifts implying expansion velocities up to 20 000 km s^{-1}.

No supernova (of either type) has been observed in our own galaxy since telescopes were invented, so that all the observations are for supernovae in other galaxies. Type I supernovae are found in all types of galaxy, and occur at an apparent rate of about one per century in the larger ones. Their presence in elliptical galaxies suggests that the masses of the objects becoming such supernovae are little more than that of the Sun, because more massive stars would have evolved to white dwarfs in such galaxies long ago.

Type I supernovae, again from their distribution in galaxies, are thought to originate from older stars. Though they are not well understood, most astronomers agree that they probably originate from close binary systems containing white dwarf stars (figure 7.3, see also Journey 6).

The initial composition of such a white dwarf after its formation would

Figure 7.2   A type I supernova in Cen A (NGC 5128). (Reproduced by permission of the European Southern Observatory.)

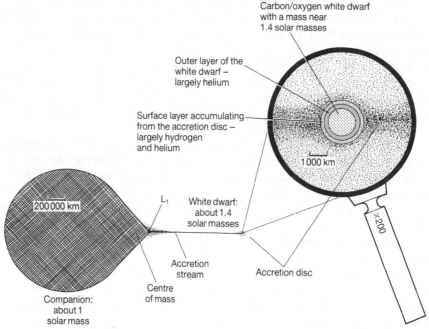

Carbon/oxygen white dwarf
with a mass near
1.4 solar masses

Outer layer of the
white dwarf –
largely helium

Surface layer accumulating
from the accretion disc –
largely hydrogen
and helium

1 000 km

200 000 km

L₁

White dwarf:
about 1.4
solar masses

×200

Accretion
stream

Accretion disc

Centre
of mass

Companion:
about 1
solar mass

Figure 7.3   Schematic cross section through a close binary system likely to become a type I supernova (cf models for novae, Journey 6).

be largely carbon or oxygen. The companion star would be losing mass and this would accumulate onto the surface of the white dwarf. However, as we saw on our previous journey, it is in just such systems that novae are thought to originate. It is possible therefore that the system may explode in one or more nova outbursts before it becomes a type I supernova.

Whether nova explosions occur or not, the accumulating material would cause a slow collapse of the whole white dwarf, and a rise in its internal temperature. Eventually the central temperature would rise sufficiently for the carbon or oxygen to be consumed in nuclear reactions. These reactions probably occur rapidly, but not explosively. Nonetheless some 10^{44} J would be released in a few tens of minutes. This amount of energy would take the Sun the whole of its main sequence lifetime to radiate away: 10^{10} years. Its release over such a short interval of time would be more than sufficient to disrupt the white dwarf completely, and to expel its material at velocities of 10 000 km s^{-1} or more.

The mass of the white dwarf prior to the explosion is estimated to be around one solar mass. Some two-thirds of this material would be synthesised into nickel-56 from the carbon and oxygen during the explosion. However, nickel-56 is radioactive and decays to cobalt-56 and then to iron-56 with half-lives of 6.1 and 77 days respectively. The energy released by these decays would provide the power for the later stages of the supernova's emissions. Thus we get the observed 'half-life' of 50 to 80 days for the decline in the supernova's brightness after the first few weeks, from the decay of cobalt to iron. Recent x-ray observations of the remnant of Kepler's supernova have shown that most of the ejected gases do indeed consist of vaporised iron, and this observation has provided the first strong support for this model of type I supernovae.

Much of the material of the white dwarf would be flung out into space as a hot gas cloud, in most cases leaving little or nothing behind. Sometimes, though, a neutron star may remain at the centre of the expanding gas shell and perhaps become a pulsar (see later in this journey).

Type II supernovae are essentially all those which are not type I! They are thus a much more varied group of objects. Their peak brightnesses are lower (figure 7.4), typically 10^6 to 10^8 times brighter than the Sun. They tend to remain at their peak brightness for longer than type I supernovae, and then to fade more slowly. The spectra of type II supernovae (figure 7.5) contain hydrogen lines as well as those of other atoms, and indicate expansion velocities up to 10 000 km s^{-1}.

Type II supernovae are observed to occur exclusively in the spiral arms of the more open spiral galaxies, where perhaps one per 50 years might be expected. Recently, however, radio astronomers have detected many supernovae which never become optically visible because of obscuration

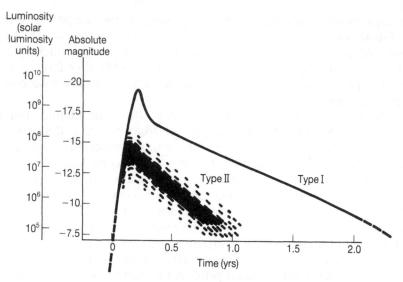

Figure 7.4   Light curves of type I and type II supernovae.

by gas and dust. The true rate is thus perhaps twice the optically observed rate, and the total of type I and II supernovae in a large spiral galaxy may reach five to ten per century. Given the duration of a supernova near its maximum, we may thus expect about one galaxy in a hundred to have an observable supernova in it at any given time.

Type II supernovae clearly originate from young stars because they have such a close association with the spiral arms of galaxies whose shapes are outlined by very bright, young stars. These supernovae are thought to be better understood than the type I's, and the neutrino pulse of the 23rd of February has gone far towards confirming those ideas.

When we followed the Sun through to its demise, we saw that the nuclear reactions involving hydrogen, which are its current energy source, will eventually be superceded by the triple alpha reactions involving helium. The result of those reactions is carbon, and perhaps another helium nucleus may then be added to form oxygen. These elements in turn, as we have just seen with the type I supernovae, may also be consumed in nuclear reactions. In fact, there is a hierarchy of reactions, starting from hydrogen, progressing through helium, carbon, oxygen, magnesium, silicon, sulphur and finally reaching iron.

Now iron is only element 26 out of the 103 known elements. So why does the sequence stop there? The answer is simple, the reactions up to iron release more energy than goes into starting them. After iron, more energy is required to build up the heavier elements than is released by

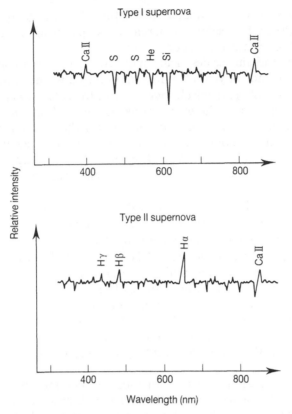

Figure 7.5   Schematic spectra of type I and type II supernovae. Note the strong hydrogen emission lines in the spectrum of the type II supernova, and the absence of hydrogen lines in the type I spectrum. Note that spectra vary markedly with time and between individual supernovae.

their formation. Thus *exothermic* or energy-releasing reactions can only occur until iron has been produced†.

The central temperatures of most stars will never rise beyond the point at which hydrogen can be burnt. Stars of the size of the Sun may progress to producing carbon and perhaps oxygen, but no further. Only the very small fraction of the stars whose masses are 10 or more times

---

† Of course, energy can be produced by the *breakdown* of heavy to light elements, so long as the final products are iron or heavier elements. It is this process which provides the energy for nuclear power stations and atom (as opposed to hydrogen) bombs.

that of the Sun will be capable of reaching internal temperatures high enough for the whole sequence of reactions up to iron to occur. It is these stars that become the type II supernovae.

Normally the timescales dealt with by astronomers are very long indeed: thousands of years for natural changes in the continents of the Earth, millions of years for changes within the solar system, and billions of years for change to the gross properties of the Sun. Only right at the start of Journey 1 did we encounter, amidst the turmoil of the start of the big bang, and amongst the novae on our last journey, major changes occurring in seconds or less. Now, however, the final stages in the life of a massive star which lead to type II supernovae occur over days or hours, and finally over a few milliseconds.

An effect which we have encountered before (Journey 3) and will encounter again is that the longer the lifetime of a particular stage of a star's lifecycle, the more stars we are likely to be able to see in that stage. Thus, solar-type stars which spend 10^{10} years with only minor external changes are plentiful. The smaller, cooler stars which spend 10 or 100 times longer in their equivalent stages form the majority of stars in the galaxy. By this token, the brief time span of a type II supernova must make them rare objects. Indeed, in a galaxy like our own Milky Way with 10^{11} stars no such events are occurring at all for most of the time. Only perhaps two or three times in a century will a type II supernova flash into being.

Let us start the main part of our journey therefore by travelling back some ten million years to when the star which became the supernova in the Large Magellanic Cloud first came into being. That star had been observed and catalogued before its explosion, and so from its cataloguer and its position it was known in the prosaic astronomical fashion as Sanduleak $-69°\,202$.

The star would have formed from a huge swirl of thin gases in the way we saw on our third journey. But right from the start it would have stood out from the crowd. Though initially still hidden by gas and dust left from its formation, it would have been brighter than all but one in 10 000 of its youthful companions. Had anyone been around to measure its mass, then they would have found it to be one of those rare stars which did not fragment into several small stars in the last moments of its condensation, and so it started life with perhaps 30 times the amount of material to be found in the Sun at present.

Like the Sun and other stars, our pre-supernova star would initially have been converting hydrogen into helium as its principal energy source. There would, however, have been two major differences from the situation in smaller stars. Firstly, convection would have been occurring throughout the centre of this large star and thoroughly mixing it up. Hence the whole of the central parts of the star would be being converted uniformly to helium. Secondly, the star would be burning the hydrogen very much more rapidly than is the case with smaller stars. So, although it had some 30 times the amount of hydrogen that was available to the Sun, it would consume most of it in about 0.1% of the time taken by the Sun: about 10^7 years.

Once the hydrogen in the central regions of the star had been converted to helium, the reactions would naturally come to an end. The centre, or core of the star, now almost pure helium, would then start to contract since the gravitational forces would no longer be balanced by the pressure. Potential energy would be released, heating both the core and the surrounding region, until hydrogen fusion recommenced in a shell around the core.

The core would continue to contract and heat up further until eventually temperatures and densities high enough for the helium to take part in nuclear reactions were achieved. As we saw on Journey 4, this requires three helium nuclei to collide with each other within about 10^{-16} seconds, and results in a carbon nucleus. For such a large star as Sanduleak $-69°\,202$, helium fusion would start at a temperature around 5×10^7 K, and when the density reached a thousand times that of water.

Once the triple-alpha reactions started, the helium would have been consumed much more rapidly than the hydrogen. In this case half a million years would suffice to use up the helium. As the temperatures and densities rose further, helium and carbon would combine to form oxygen, neon, magnesium and silicon. The core resulting from the fusion of helium would thus have been a mixture of several elements.

After the helium had been burnt, the core would again shrink, releasing potential energy, and raising the temperature further. Eventually, at about a million times the density of water and at several hundred million degrees, the carbon nuclei would start to react with each other, producing magnesium. At somewhat higher temperatures and densities, neon, oxygen, sulphur and finally silicon would also react. The carbon burning period would have lasted for less than a thousand years, neon and oxygen for less than a year, and the silicon burning would have been over in a day or less.

The inner structure of a large star at this stage can become complex, with an onion-like structure (figure 7.6). Each of the reactions can still be occurring in a layer at some point inside the star with the hydrogen fusion outermost, and the silicon fusion at the centre. At a given instant, just one of the reactions is likely to be dominating the energy generation, but the precise details of this are beyond the capacity of present-day computers to determine.

Now, the last fusion reactions largely result in iron-56, the most common isotope of iron, and its production marks the end of the line. For, as we have already seen, the build-up of heavier elements from iron

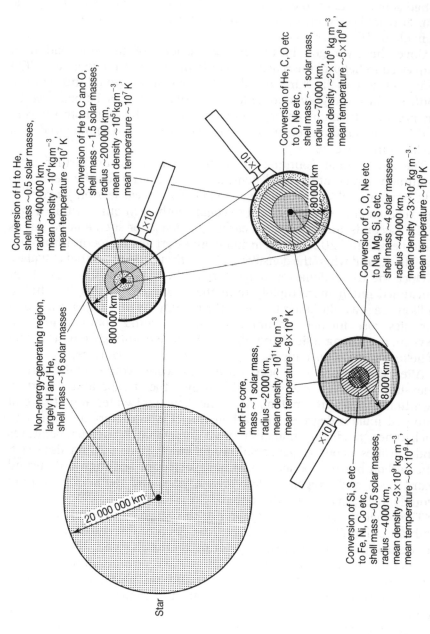

Conversion of H to He, shell mass ~0.5 solar masses, radius ~400000 km, mean density ~10^4 kg m^{-3}, mean temperature ~10^7 K

Conversion of He to C and O, shell mass ~1.5 solar masses, radius ~200000 km, mean density ~10^5 kg m^{-3}, mean temperature ~10^7 K

Non-energy-generating region, largely H and He, shell mass ~16 solar masses

Conversion of He, C, O etc to O, Ne etc, shell mass ~1 solar mass, radius ~70000 km, mean density ~2×10^6 kg m^{-3}, mean temperature ~5×10^8 K

Conversion of C, O, Ne etc to Na, Mg, Si, S etc, shell mass ~4 solar masses, radius ~40000 km, mean density ~3×10^7 kg m^{-3}, mean temperature ~10^9 K

Inert Fe core, mass ~1 solar mass, radius ~2000 km, mean density ~10^{11} kg m^{-3}, mean temperature ~8×10^9 K

Conversion of Si, S etc to Fe, Ni, Co etc, shell mass ~0.5 solar masses, radius ~4000 km, mean density ~3×10^9 kg m^{-3}, mean temperature ~6×10^9 K

Star

Figure 7.6   The probable internal structure of a 20 to 30 solar mass star a few hours before it becomes a type II supernova.

is endothermic: it requires more energy to be put into the reaction than comes out of it at the end.

The final stage in which the iron is produced (figure 7.7) occurs very rapidly and is brought to an end not by the depletion of the lighter elements, but by the collapse of the core, which occurs because the iron core no longer has any significant energy sources within it. The core resists its own gravitational collapse and the weight of the overlying layers of the star through the pressure of degenerate electrons. This is the same situation that faces white dwarfs and which we have encountered on our previous three journeys. Just as for an isolated star therefore, there is an upper limit to the mass that can be contained within the core if it is to remain stable. For an isolated white dwarf that limit, the Chandrasekhar limit, was just under 1.5 times the mass of the Sun. For the conditions experienced by the iron core, the limit is less certain, but is probably just over one solar mass (figure 7.8).

Let us therefore take up the story again when the iron core of our pre-supernova star reached its limiting mass. As soon as this happened, the core would have collapsed rapidly (figure 7.9), halving its size every few tens of milliseconds.

Previously, shrinkage of the centre of the star has led to the release of potential energy and to the raising of its temperature and pressure. Thus one might expect a similar effect this time, and for the resulting increase in pressure to slow down the collapse. In fact, though the temperature in the core would indeed have risen, the total pressure would actually fall, and thus allow the collapse to accelerate. The reason for this apparent paradox is that the temperature by this time was higher than has been found anywhere else in the universe since the first second of the big bang: perhaps 10^{10} K. The photons, in the form of high energy gamma rays, would be able to break up some of the iron nuclei into smaller particles (figure 7.10). This of course would raise the pressure arising from the nuclei, because their numbers would be increased. But the break-up of the iron nuclei would require the input of a large amount of energy. This energy could only come from the electrons, and so would lead to a decrease in the electron pressure. The decrease in the electron pressure would far outweigh the increase in the pressure of the nuclei, and so the total pressure would fall. The pressure in the core would have dropped even further as the density rose towards 10^{14} $kg\,m^{-3}$. This time the fall would occur through the loss of electrons as they combined with protons to form neutrons.

The formation of the neutrons, together with many of the other reactions occurring at the time, would release neutrinos, and these could escape from the core (figure 7.11). The neutrinos would carry off energy, cooling the core, and allowing the implosion to continue to accelerate. It was just such a burst of neutrinos that was detected for Sanduleak

$-69°\ 202$ by the nuclear physicists on the 23rd of February.

The initial collapse would have continued until the core was reduced from a radius of a few thousand kilometres to a hundred kilometres or so. The end of the first stage of the collapse would happen as a fundamental change occurred in the physics of the situation. When the density rose to about 4×10^{14} kg m^{-3} (figure 7.12) the neutrinos could no longer escape without hindrance, but would become trapped within the core. The neutrinos would eventually escape of course, but only after many scatterings, absorptions and re-emissions, and these interactions would delay them for longer than the remaining stages of the collapse. The energy loss from the collapsing core would therefore effectively be reduced to zero. The neutrinos would probably become the most significant source of pressure in the inner core and this would slow down, but not halt, the collapse.

The collapse of the core would have come to an end only when the density reached that found in atomic nuclei: 2 to 3×10^{17} kg m^{-3}. The centre of the core would then effectively become a single gigantic nucleus, and the developing nucleon pressure would be sufficient to bring the collapsing central parts of the core to a sudden halt (figure 7.13). The infalling material would overshoot its equilibrium point before stopping, resulting in a maximum density perhaps 50% too high for stability. The core of nuclear matter would therefore immediately rebound outwards, and then reverberate around its equilibrium density with a period of about a millisecond.

Now the material outside the central core would still be falling inwards (figure 7.14). The halt of the central collapse and its rebound would have generated pressure waves (sound waves) containing an appreciable fraction of the energy that had been released. These pressure waves would then have moved outwards at the local speed of sound. However, the speed of sound decreases away from the centre of the star, while the velocity of the infalling material increases. Thus there would come a point, the sonic point, at which the speed of sound would equal that of the material, and the pressure waves would make no further outward progress. This point probably occurred at a distance of 50 km or so from the centre of the star.

The sound wave would become a shock wave as more and more energy from the halt, its rebound and the impact of the still-imploding outer parts of the core with the nearly-stationary central region, accumulated at the sonic point. Now shock waves travel faster than the speed of sound, and so the pressure waves could once again move outwards.

Shock waves, however, are very violent processes, and these would have been sufficiently energetic to break up the large nuclei in the material they passed through. This would have lowered both the temperature and the pressure of the material because vast quantities of energy would be consumed in the reactions. Neutrinos would also be produced as protons and electrons combined into neutrons. These neutrinos would escape since the density of the material at this distance from the centre of the star would be too low to trap them. Thus even more energy would be removed from the shock wave region. These various processes would result in the shock wave coming to a halt at only 100–200 km from the centre of the star (figure 7.15).

The way in which the supernova would then progress seems to depend upon the mass of the original star. The smallest stars which may become supernovae are probably around 8 to 10 times the mass of the Sun. These supernovae are not well understood, but it is possible that the shock wave simply has sufficient energy to disrupt the star without becoming stalled. In this it may be aided by the reactions in the core not having proceeded all the way to iron. The core therefore still contains appreciable amounts of oxygen and silicon and reactions occurring in these lighter elements will add energy to the shock wave. The supernova of 1054 which produced the Crab nebula may have been the result of one of these low mass explosions.

With stars around 15 times the solar mass, the rebound and reverberation of the central core probably inject enough energy into the shock waves for all the iron in the outer parts of the core eventually to be fragmented into protons and light elements. The shock wave then escapes into the outer parts of the star, and again results in its disruption.

With the largest stars, 25 solar masses or more, such as may have resulted in the LMC supernova, there is too much iron in the outer part of the core for the shock wave to fragment entirely into lighter nuclei. After ten milliseconds or so the shock wave therefore becomes stalled at a distance of one to two hundred kilometres from the centre.

However, an eternity later (by the standards of the speed of the changes we have just been examining): about one second, the shock wave is enabled to continue outwards. It does this through the absorption of energy from the neutrinos streaming outwards from the central core (figure 7.16).

Some 10^{46} J goes into the neutrinos, the energy that would be emitted by the Sun over 10^{12} years were it capable of surviving that long. Although we have spoken of the neutrinos being trapped inside the inner core, this is only on the millisecond timescales which have been of concern during most of the collapse and rebound. In fact, they will be lost eventually. At 150 km from the centre, perhaps only one in a thousand of these escaping neutrinos will interact with the material in the shock wave, but they carry so much energy that this is sufficient to disrupt the remaining iron nuclei and to heat the shock wave until it moves outwards again (figure 7.17). Then as before, the outer parts of the star

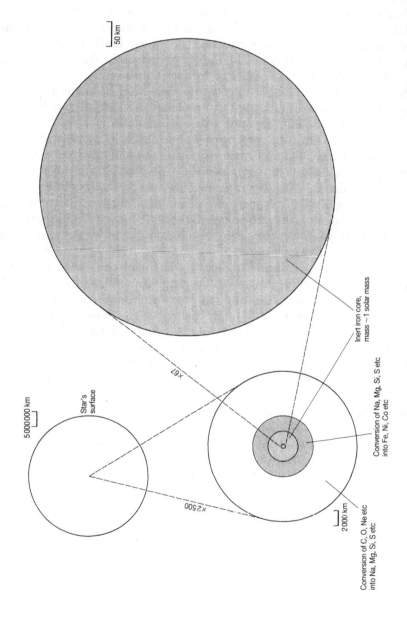

Figure 7.7   The internal processes in a 20 to 30 solar mass star undergoing a supernova implosion/explosion. All quoted times are approximate. Time $t = T_0 - 6$ hours: the formation of the iron core.

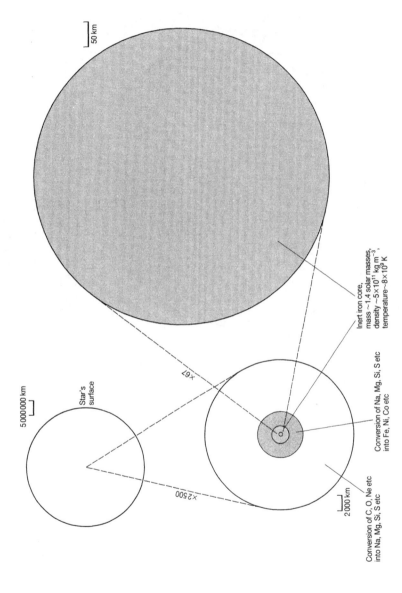

50 km

5 000 000 km

Star's surface

×67

×2 500

2 000 km

Conversion of C, O, Ne etc
into Na, Mg, Si, S etc

Conversion of Na, Mg, Si, S etc
into Fe, Ni, Co etc

Inert iron core,
mass ~1.4 solar masses,
density ~5×10^{11} kg m^{-3},
temperature ~8×10^{9} K

Figure 7.8   Time $t = T_0 - 1$ s: almost up to the Chandrasekhar limit.

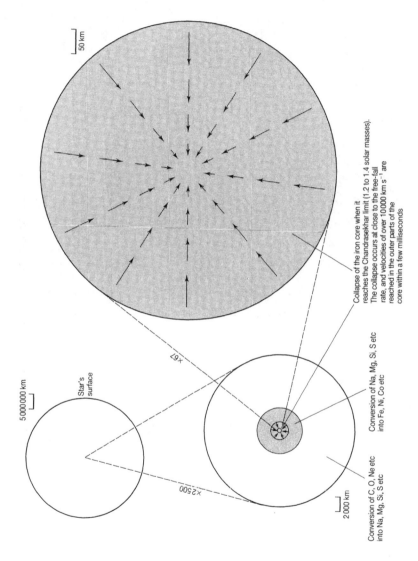

Figure 7.9   Time $t = T_0$; the start of the core collapse.

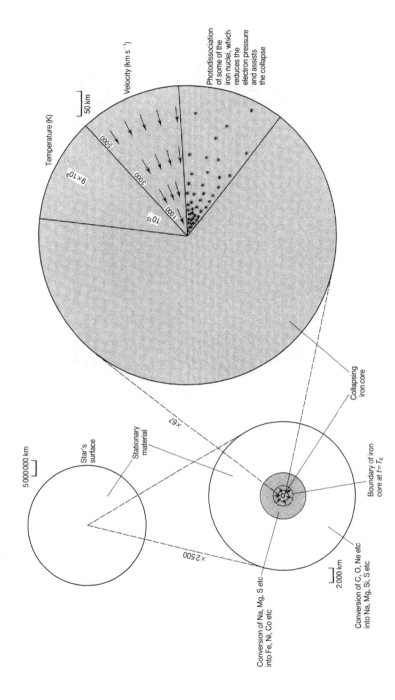

Figure 7.10   Time $t = T_0 + 30$ ms: the core collapse and the start of the break-up of the iron nuclei.

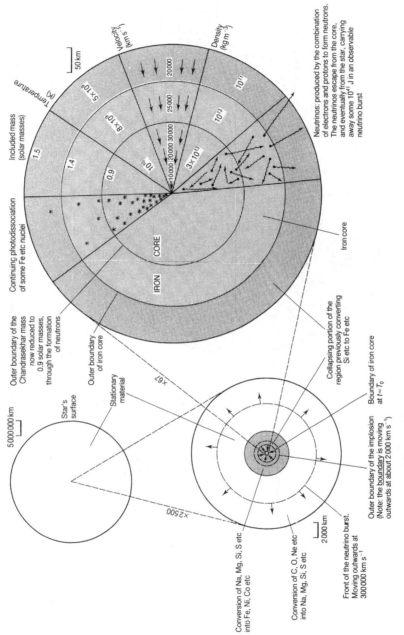

Figure 7.11   Time $t = T_0 + 75$ ms: the neutrino pulse.

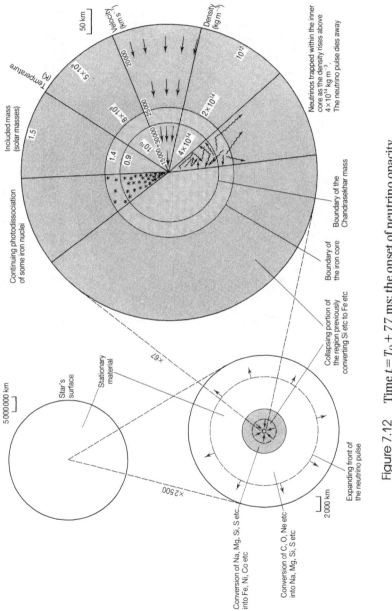

Figure 7.12   Time $t = T_0 + 77$ ms: the onset of neutrino opacity.

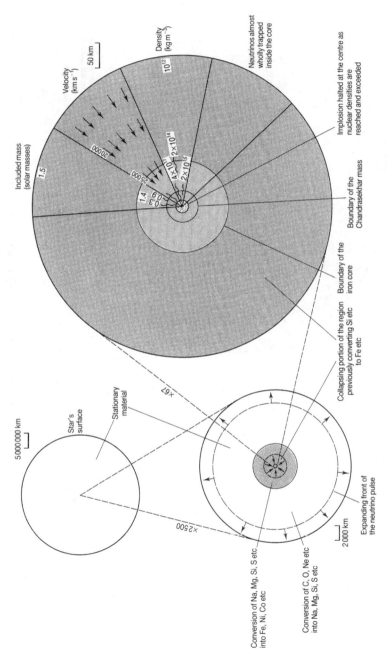

Figure 7.13   Time $t = T_0 + 80$ ms: maximum density, the collapse starts to halt.

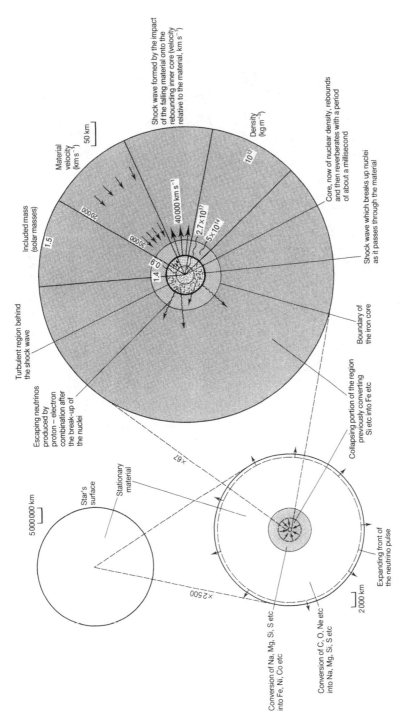

Figure 7.14   Time $t = T_0 + 82$ ms: the core rebound and the formation of the shock wave.

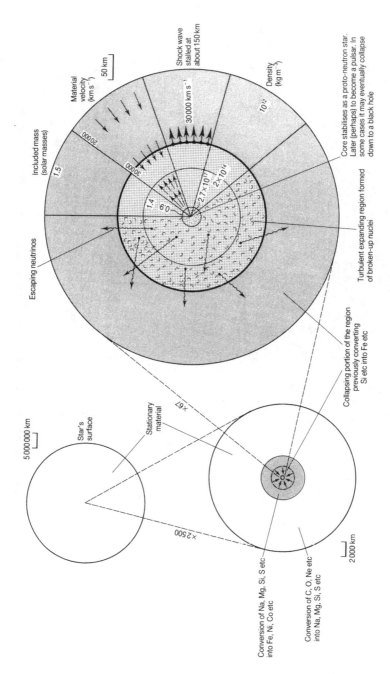

Figure 7.15 Time $t = T_0 + 100$ ms: the shock wave stalls.

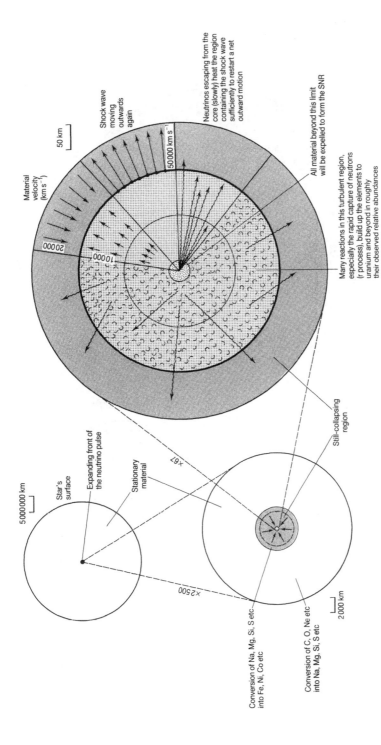

Figure 7.16   Time $t = T_0 + 1$ s: the shock wave revivified.

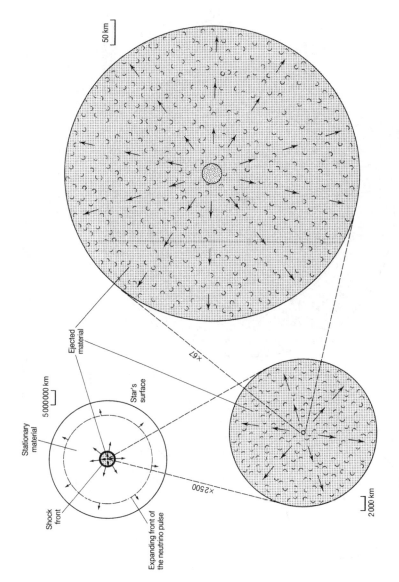

Figure 7.17   Time $t = T_0 + 50$ s: the shock wave progresses outwards.

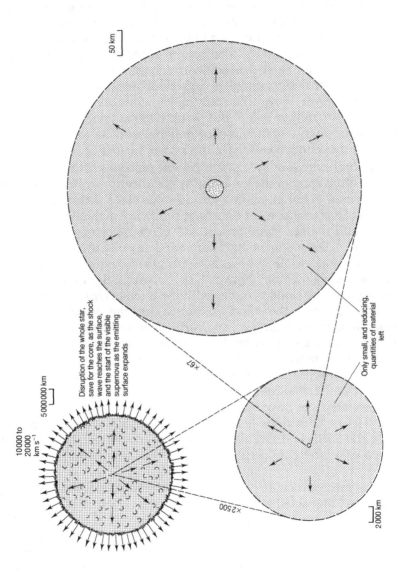

Disruption of the whole star, save for the core, as the shock wave reaches the surface, and the start of the visible supernova as the emitting surface expands

Only small, and reducing, quantities of material left

50 km

5 000 000 km

10 000 to 20 000 km s⁻¹

×67

×2 500

2 000 km

Figure 7.18   Time $t = T_0 + 30$ min: the disruption of the star and the start of the optical phase of the supernova.

explode outwards into the visible supernova (figure 7.18). Upwards of 20 solar masses of material may then be expelled in the final convulsion.

In most cases it is the sudden brightening of the star which announces the occurrence of the supernova to human astronomers. The initial implosion has been halted by the extreme rigidity of the material in the core as it reaches nuclear densities and is converted into a disrupting explosion of most of the outer parts of the star. The expansion of the surface area which this causes then results in the initial increase in the brightness of the star. Later the radiation may result from other sources such as radioactive decay, or high energy electrons from the remnants of the core. The total energy emitted in the form of electromagnetic radiation may amount to 10^{44} J. But perhaps ten times this amount goes into the kinetic energy of the expelled material and, as we have seen, a hundred times as much is emitted in the form of neutrinos.

After all this hectic activity, the subsequent processes slow down to the sedate pace more appropriate for astrophysics. Clearly we are left with an expanding cloud of hot gases, and also in some cases with a small, very dense lump of material at what used to be the centre of the star.

To deal with the gas cloud first, it will continue to expand at about $10\,000$ km s^{-1}. Depending on its distance and any obscuring dust, it may eventually become visible to us as a beautiful supernova remnant (SNR) such as the Crab nebula (figure 7.19). Continuing expansion will eventually cause the cloud to merge with the interstellar medium. There, the cloud will have the effect of increasing the temperature and turbulence of the interstellar gas, and perhaps in some cases of compressing another nearby gas cloud until star formation occurs (Journey 3).

More importantly from our point of view, the abundance of elements heavier than hydrogen and helium in the interstellar medium is enriched by an influx of the massive nuclei produced during the supernova. The solar system seems to have been influenced by at least two supernovae (Journey 3). One formed the heavy elements which make up the planets including the Earth and ultimately, of course, ourselves. Then another supernova at a later date perhaps initiated the actual formation of the Sun.

In many cases, particularly with the type I supernovae, the expanding gas cloud is all that is left after the explosion. But in other cases, there is also a highly compressed, compact object arising from the collapse of the core. As we have seen, this is of nuclear density. Most of the protons and electrons in it will have combined into neutrons, and so it is known as a neutron star. If its mass is over about twice that of the Sun, then as it cools even nuclear forces will be unable to generate enough pressure to balance gravity, and it will collapse again—this time down to a black hole (see Journey 8). If the mass is less than about 1.5 solar masses, then

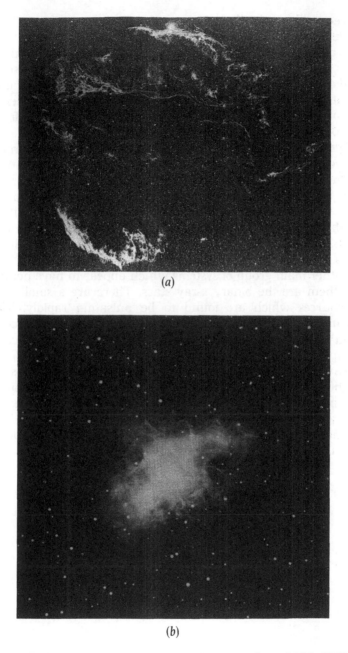

(a)

(b)

Figure 7.19 Supernova remnants. (a) Cygnus loop (NGC 6960–62). (Reproduced by permission of the Hale Observatory.) (b) Crab nebula (NGC 1952, M1). (Reproduced by permission of the Royal Astronomical Society.)

the object will eventually stabilise with a radius of 10 to 20 km and a mean density of about 10^{17} kg m^{-3}.

Several hundred such neutron stars have been observed in the last two decades, most of them occurring as *pulsars*. These are usually observed in the radio region of the spectrum, and the name derives from their main observational characteristic: **pulsa**ting **r**adio **s**ource. They emit sharp pulses of radio waves at highly regular intervals ranging from a few milliseconds to a few seconds.

The precise cause of the pulsar's radio emission remains a problem, but the periodicity is just due to the rotation of the neutron star. Like a lighthouse, a bright flash is seen once per rotation when the emitted beam is along our line of sight. However, only a very few pulsars have actually been found inside gaseous supernova remnants, while the processes that we have just reviewed for type II supernovae would suggest that most SNRs should have pulsars. The reason for this discrepancy remains something of a mystery.

The other main group of systems which appear to have neutron stars within them are the binary x-ray stars. There are a small number of x-ray sources which are found to be pulsating rapidly. The latest explanation for them envisages a binary system which contains a neutron star. The x-ray emission arises as material from the companion falls onto the surface of the neutron star.

Thus we come to the end of our seventh journey. It has taken us through the hottest spots still to be found in the universe, shown where the elements which go to make up our bodies come from, and into regions as dense as the nucleus of an atom, but containing 10^{57} times as much material. Only on our first journey and on our next journey will we encounter greater densities.

Journey 8

Cosmic Dustbins

". . . I wish you wouldn't keep appearing and vanishing so
suddenly: you make one quite giddy."
"All right" said the cat; and this time it vanished quite slowly,
beginning with the end of the tail, and ending with the grin which
remained some time after the rest of it had gone.

Alice's Adventures in Wonderland
L Carroll

Black holes have achieved a widespread public notoriety occasioned by numerous 'sci-fi' movies wherein the gallant crew, usually led by a buxom and sparsely clad female captain, battle valiantly to repair their ship's engines. Inevitably they succeed just in time to avoid being sucked to their deaths into one of these dustbins of the universe.

Such public acclaim therefore makes it surprising that outside the film studio, no black hole has yet been found with certainty. Furthermore, theory suggests that black holes would actually take an infinitely long time to come into existence. Thus in a universe that is less than infinitely old, no black holes can yet have formed anywhere.

The 'dustbin' aspect of black holes: their apparent ability to engulf any passing item that strays too close and never to allow it out again, is the one which has caught most people's imagination. However, the actual diet of black holes must be expected to be much less exotic than distressed spacecraft and their amatorial crews. Usually, individual atoms and molecules from the gas between the stars, flavoured by rare microscopic dust particles, will be the main items for consumption. Some more enterprising black holes though, may be chewing up nearby stars or even parts of galaxies.

The dustbin aspect is only one, and by no means the most interesting, of the predicted properties of black holes. To discover some of the others we will have to travel beyond the limits wherein astronomy and physics are understood with any certainty. Inevitably, given the lack of known black holes, much of our journey must be through a theoretician's paradise.

Let us make a start on our journey by looking at the basic problem of 'What is a black hole?'

The idea of a black hole appears at first sight very straightforward. It is well known that Einstein, early in this century, showed that no material object could travel as fast as the speed of light in a vacuum ($299\,793$ km s^{-1}). It is almost equally well known that there is a minimum velocity, the escape velocity, which a rocket has to achieve if it is to leave the Earth's gravitational influence. For the Earth, that velocity is 11.2 km s^{-1}†, for our Moon it is only 2.4 km s^{-1}, while for the Sun it is 617.7 km s^{-1}. Clearly for more massive stars, the escape velocity could be even higher, 930 km s^{-1} for Spica (figure 8.1) for example.

However, the escape velocity depends not only on the mass of the object, but also upon its size. Thus a white dwarf (Journey 5) with a mass equal to that of the Sun would have an escape velocity ten times that of the Sun, or nearly $7\,000$ km s^{-1}, whilst for a solar mass neutron star (Journey 7) the escape velocity would be $120\,000$ km s^{-1}. We can extend this sequence in our imaginations until we have an object whose escape velocity is equal to the velocity of light. For an object with the mass of the Sun, this equality would occur when it had condensed down to a radius of about 2.6 km. Since no material substance can travel at the speed of light, nothing would then be able to escape from that condensed object. Material though, would still be able to fall into the black hole. Light and other electromagnetic radiation such as radio waves, x-rays etc, would be trapped after only a very slight further decrease in the object's size.

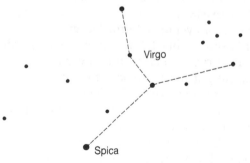

Figure 8.1   Spica: a star with a mass 15 times that of the Sun.

---

† Since most spacecraft are launched into low orbits, they do not completely escape from the Earth's gravitational field, and thus require launch velocities of 'only' between 7.8 and 9 km s^{-1}.

This simple picture of why black holes exist is quite wrong. It is perhaps unfortunate, therefore, that the simple approach does result in the correct answers for non-rotating black holes. A solar mass black hole does indeed have a radius of about 2.6 km. The reason that this simple picture is incorrect is that the idea of an escape velocity is rooted in Newton's theory of gravity, while the idea of the speed of light as a limit is a result of special relativity, and is not a part of Newtonian physics.

Only by using a theory of gravity which incorporates the speed of light as an upper limit to the transmission of all signals, including changes in gravitational effects, can a reliable picture of the properties of black holes be obtained. The main (but not the only) contender for such a theory at the moment is, of course, general relativity. While full understanding of general relativity requires a very high level of mathematical competance, some of its underlying concepts, and many of its predictions, can be appreciated without the use of mathematics.

In general relativity then, how is the idea of a black hole formulated? To see this, we must first appreciate the fundamental difference between the way in which gravity is viewed in Newtonian mechanics and in relativity. The Newtonian view of gravity is as a force field which acts instantaneously at any distance from the mass producing it. In relativity, however, gravity is viewed not as a force at all, but only an apparent force which results from the incomplete view of the circumstances taken by the observer†. The effect of a mass, according to general relativity, is to distort the fabric of space-time (figure 8.2). Changes in this distortion of space-time due to changes in the mass then propagate outwards in the form of gravitational waves at the speed of light. Objects moving within the distorted space-time fabric, if unaffected by other forces, do so along paths which represent the shortest distances between successive points (geodesics). In our three-dimensional view of the world, such paths are seen as curved, leading to the orbital motions of the planets and all the other effects we normally attribute to the 'force' of gravity.

On the general relativistic view of gravity then, the space-time around a black hole differs from that around less concentrated masses only in having a greater degree of distortion. The paths of moving objects coming from the black hole are so curved that the objects will turn back on themselves and their paths re-intersect the surface of the black hole.

It may seem that general relativity simply provides an alternative description of escape velocity. In fact, the different approach taken by general relativity to describing black holes has major implications for the consequent predictions of the properties of the black hole.

---

† An equivalent problem occurs with centrifugal 'force'. It certainly feels like a force to a rider on a big dipper, but in fact it is an acceleration arising from the application of the centripetal force in the opposite direction.

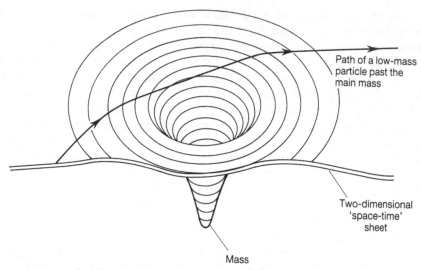

Figure 8.2   Schematic representation as a 2D sheet of the distortion induced in the 4D fabric of space-time by the presence of a mass.

As one immediate example, consider a photon originating at the surface of the black hole and moving outwards. Now in relativity, the local velocity of a photon is always the velocity of light, but the escape velocity reduces as we move outwards. The photon therefore, some distance above the surface of the black hole, *will* have a velocity exceeding the escape velocity from that point. It might therefore appear that photons would still be able to escape from a black hole.

However, relativity also predicts that a photon originating in a gravitational well and observed from outside that well will be red-shifted. That is to say, the photon will have its wavelength increased and its energy reduced compared with the situation had it originated further out. In effect the photon loses energy as it climbs the 'hill' to get out of the gravitational well. The loss of energy by this process is such that a photon starting from the surface of the black hole would have lost all its energy by the time it had escaped. In other words, the photon's wavelength would have become infinite and its frequency zero, it would carry no energy or information, and we would be unable to observe it. For all practicable purposes then, photons as well as particles are predicted by general relativity to be unable to escape from a black hole.

One warning flag, however, needs to be raised at this stage. So far physicists have only been able to check the validity of general relativity in weak gravitational fields. It may be that general relativity's inevitable approximation to the truth will become apparent in the intense gravitational fields of a forming black hole. The predictions of general relativity will break down and it will have to be replaced by a better theory, just as general relativity replaces Newtonian theory when that becomes inaccurate. Such a new theory may, of course, predict very different properties for black holes.

Bearing this last caveat in mind, we find that the Newtonian and relativistic approaches produce identical results for non-rotating, uncharged black holes (often called Schwarzschild black holes). The radius of such a black hole, the Schwarzschild radius, is not that of a solid object. It is the radius of the volume within which, to use the Newtonian description, the escape velocity exceeds the speed of light. The 'surface' of a black hole is thus called the event horizon to indicate its non-solid nature, and because it is the last point from which the outside world can observe events occurring near a black hole.

Material falling into a black hole is predicted to be crushed into a mathematical point, of zero physical dimensions, at its centre (a phenomenon usually called a singularity and which cannot be described by today's laws of physics).

Apart from mass, rotation and electric charge are the only properties of the material forming a black hole that may be detected after its formation. In practice, most black holes may be expected to possess considerable angular momentum, but probably little net electric charge. Such black holes are known as Kerr black holes, and we shall return to their properties later.

Sticking with the simpler Schwarzschild black holes for the moment, let us see what differences the relativistic approach to their description gives us. One of the most fundamental differences from the Newtonian predictions has already been mentioned. It is that black holes will actually never form in practice, because that would require an infinite length of time, though Newtonian theory suggests that they should form very quickly indeed. Thus, according to Newtonian physics, a neutron star pushed over its mass limit by, say, acquiring matter from its companion star in a binary system, should collapse to a black hole in about a millisecond.

Relativity predicts much the same behaviour patterns as Newtonian theory until the material has collapsed to within a metre or so of forming the black hole. Then, however, an effect unknown in Newtonian theory starts to become significant. This effect is gravitational time dilation†,

---

† One of the better known predictions of special relativity is that clocks slow down on an object that is moving very rapidly, compared with clocks that are stationary with respect to the observer. It is less well known that a similar effect is predicted for clocks operating within a strong gravity well compared with those further out: gravitational time dilation. Surprising though these effects appear by everyday standards, they have both been confirmed by experiment.

and it leads to the previously accelerating infall slowing and eventually coming to a halt, *as seen by an outside observer*.

Thus if we imagine ourselves in the vicinity of a black hole, but sufficiently far away to be outside its significant gravitational effects, then we could watch as a brave (or idiotic) volunteer descended towards the hole carrying a clock. Comparing that clock with a similar one of our own, we would notice that as the volunteer got deeper into the gravitational field of the black hole his or her clock would slow down.

When the volunteer was one Schwarzschild radius out from the event horizon (2.6 km for a solar mass black hole), one second on his or her clock would last 1.4 seconds according to our clock. At a tenth of a Schwarzschild radius from the event horizon, one second would be lasting 3.3 seconds, while when the volunteer reached one thousandth of a Schwarzschild radius from the black hole his or her seconds would be lasting 32 seconds by our clock. At the event horizon itself, we would find that one second on the volunteer's clock lasted an infinite length of time according to ours, i.e. the volunteer's clock would appear to have stopped.

Of course, it is not just clocks which are affected by this gravitational time dilation, all physical events will similarly slow down. In particular, the motion of the volunteer will be affected. If he or she were simply falling into the black hole, then we would initially see a rapidly accelerating inward velocity, but as the volunteer got to within a few metres of the event horizon, the effects of gravitational time dilation would start to overcome those of the increase in velocity. To observers on the *outside* then, the volunteer's fall would appear to slow down rapidly, and after an infinite length of time, he or she would come to a halt at the event horizon itself. The same considerations would apply to any object falling into a black hole and even to a collapsing object forming a new black hole. Thus, given the observed age of the visible universe of about 15 aeons (a long time, but far short of infinity), there will not have been sufficient time for any black holes to have formed anywhere.

Now in some ways, claiming that no black holes can yet have formed in our universe is a case of academic hair-splitting. An object collapsing to form a black hole would shrink to within a micron of the event horizon within a few seconds, even as seen from the outside. At that stage it would have most of the properties of a genuine black hole and for most practical purposes could be treated as such. But there is one important difference—the material does not pass through the event horizon and go on to form a singularity. Thus, at least for the rest of the universe, that unpleasant phenomenon is avoided.

Now so far, we have looked at black holes from the point of view of an outsider. What would our intrepid volunteer have seen? The answer, had he or she been unwise enough to choose a black hole containing only one solar mass, would be 'not a lot'. Tidal effects (the differential gravitational forces between the volunteer's feet and head) would have converted the volunteer into minced meat long before the astronomically interesting effects started happening.

However, tidal effects become smaller as the black hole becomes more massive, even though the total gravitational field is increasing. Thus if the volunteer were to choose a black hole with the mass of a galaxy: 10^{10} to 10^{11} solar masses, or the sort of black hole that might perhaps be found at the heart of a quasar, then the forces trying to pull him or her apart would only reach a few hundred newtons even at the event horizon. Such a force is only about the equivalent of swinging by the arms from a circus trapeze, and so the volunteer would be able to survive their effects.

A black hole with the mass of a galaxy would, of course, be much larger than one of one solar mass: up to 2.6×10^{11} m, or about the size of Mars' orbit. Falling in from a distance would therefore take much longer, and the volunteer would not only survive the tides at least as far as the event horizon, but would have time to look around and see what was happening.

The volunteer's account of his or her adventures would differ radically from that of the outside observer. There would not be any slowing down as the event horizon was approached—the velocity would just continue increasing. He or she would pass through the event horizon at the speed of light, ending up a few minutes later crushed into the singularity. The tides, of course, continue to increase inside the event horizon, so even though the volunteer may have survived the passage through the event horizon, he or she would be torn assunder soon afterwards.

How can we reconcile two such apparently conflicting views? In one case the volunteer passes through the event horizon, in the other he or she comes to a halt at it.

The answer, as in other apparent paradoxes of relativity, is that there is no true paradox unless two observers *with identical points of view* see things differently. If there is any difference between the observers: position, time, velocity, acceleration, or in this case gravitational field, then there is no reason to expect their views of events to coincide†.

We may be able to reduce the force of the apparent paradox somewhat by widening our consideration of the volunteer's view of events. Let us suppose that the volunteer chooses to look backwards, perhaps to wave a last fond farewell. Then he or she would see our clock going faster and

---

† As a much simpler, non-relativistic example of this effect, consider two supernovae (figure 8.3) in a galaxy: an observer at A sees them occurring simultaneously, while an observer at point B would see one supernova occurring 100 000 years or so before the other.

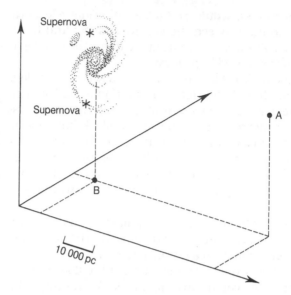

Figure 8.3   An example of position affecting views of events.

faster as he or she fell closer and closer to the black hole. As the event horizon approached, our clock would speed up enormously. As the volunteer reached the event horizon, the whole lifetime of the outside universe would pass by in an instant and it would come to its end. There would thus no longer be an outside observer to have a viewpoint to differ from that of the volunteer!

Now in case any soft-hearted readers are concerned at this lonely and cruel fate awaiting our volunteer, let them be reassured that he or she will know nothing of such things in practice. On a clear dark night, if we go outside, we will be bathed in starlight amounting to about 20×10^{-6} W m^{-2}. That is enough energy to raise our temperatures by about three millionths of a degree. This same star light will also illuminate our volunteer. However, as he or she approaches the event horizon, that light will be blue-shifted, firstly towards the ultraviolet region, then into x-rays, and finally into gamma rays. Additionally its intensity will increase because in one of the volunteer's seconds, many seconds, or centuries, or eventually aeons, will pass in the outside universe.

By the time that one century passes in the outside world for each second to the volunteer, the star light, aided by cosmic rays and the microwave background radiation will have killed the volunteer by radiation poisoning and then evaporated his or her remains. So all that will survive of our volunteer, the clock and any spacecraft he or she may have used to reach the event horizon, will be a glowing cloud of ionised and radioactive gases.

The idea that has been put forward that black holes could form gateways into other parts of the universe or even to other universes, a concept sometimes called wormholes in space, will thus be of very little practical interest as a means of touring the universe.

Leaving behind the problems of forming or falling into black holes, the one property that anyone who has heard of black holes 'knows' about them, is that nothing can ever escape. Yet even that truth has been suspect for some time. It is true that an object (our volunteer for example) will not escape in a recognisable and identifiable form, but none-the-less, it is possible for a black hole to lose energy and mass, and eventually to cease to exist.

This unexpected phenomenon arises from those odd ephemera encountered on the first journey: the virtual particles. As we saw then, these come into being as pairs of particles and disappear again after a very brief instant throughout all of space. We may therefore expect them to be appearing near to black holes. Close to a black hole, however, the differential gravitational field (tides) may separate the two particles during their fleeting existence sufficiently for their recombination to become impossible. One of the particles may then eventually fall into the black hole, while the other escapes into the surrounding universe.

Now the virtual particles exist only because their lifetimes are too short for the law of conservation of mass and energy to be violated. If a pair of such particles do not annihilate each other in the Heisenberg time, then the energy 'debt' due to their formation is not 'repaid' by their destruction. The energy for their formation must therefore be found from elsewhere, and that energy source is the black hole.

The possibility of black holes acting as energy *sources* is not quite such a contradiction of all that has gone before as it may appear. We do not have an actual particle coming out through the event horizon into the rest of the universe. The virtual particles, until the energy is found to give them permanent existence, represent negative energy. The process is thus that of the inflow of negative energy (or mass) into the black hole rather than an outflow of real energy (or mass) from it.

The net effect of the virtual particles, however, is the same as if a real particle had managed to escape from the black hole. The mass of the black hole is reduced, and a new particle (the other half of the virtual pair) comes into permanent existence in the rest of the universe. The possibility of the occurrence of this process was first deduced by Stephen Hawking. The resulting emission of energy from a black hole is therefore known as Hawking radiation, because it is pairs of virtual photons (rather than particles) that will usually be involved.

The emission of the Hawking radiation by black holes means that they are no longer truly black. That is to say, a black hole's temperature is above zero degrees. However, for the sort of black holes that we may be expecting to form nowadays, say by the collapse of a neutron star, the resulting temperature is very low indeed: 10^{-8} K for a black hole with six times the mass of the Sun. Since such a black hole is bathed by the microwave background radiation at a temperature of 2.7 K, not to mention star light, cosmic rays etc, overall it will still be gaining mass.

Not until the universe reaches an age of 10^{27} years (it is currently 'only' about 1.5×10^{11} years old) will the temperature have fallen low enough for solar mass black holes actually to start losing mass. It would then take about 10^{65} years for the black hole to lose all its mass and to disappear.

However, the temperature of a black hole due to its Hawking radiation increases as the black hole gets smaller, because the radiation arises through differential gravitational effects (tides), not from the total strength of the gravitational field. Thus at a mass of about 4×10^{22} kg (half the mass of the Moon) or less, black holes would be able to lose energy faster than they acquire it from the microwave background even today. Such a small black hole could not currently be formed by any known process, but it is possible that they might have been produced in the early stages of the big bang (Journey 1).

Since the temperature of a black hole rises as it gets smaller, theoretically reaching infinity as the black hole disappears, very small black holes may be expected to be bright gamma ray sources[†]. Black holes with original masses of about 10^{11} kg (the mass of an asteroid about a third of a kilometre across) produced in the early stages of the big bang should just be reaching their final stages now. Thus one way of trying to find black holes observationally, or in this case to find where they have been, is to search for their dying bursts of gamma rays.

As mentioned earlier, the only properties of an object falling into a black hole which remain detectable afterwards are its mass, its angular momentum (rotation) and its electric charge. Now electric charges rarely build up in astronomical objects because the ionised gas or plasma that forms stars, hot gas clouds etc has a very high electrical conductivity. Thus currents can flow and neutralise any significant electrical charge that may start to form. But rotation is very common; indeed non-rotating objects are the rare exception.

We must thus expect any actual black holes we may find not to be the simple non-rotating Schwarzschild black holes so far considered, but to be spinning very rapidly through the conservation of angular momentum (figure 3.4). Such black holes are known as Kerr black holes and

---

† Recent work though, suggests that quantum mechanical effects may act to prevent the complete evaporation of black holes.

have some significantly different properties from Schwarzschild black holes.

Now the rotation of any object has the effect of dragging other, nearby, objects around. Normally the effect is so small that it is undetectable, but with the concentration of mass and the high spin of Kerr black holes it becomes significant.

Thus a space traveller close to a rotating black hole would find that the rest of the universe appeared to be spinning around. In order to appear at rest with respect to the rest of the universe, the space traveller would therefore have to spin in the opposite direction, and with increasing rapidity the closer he, she or it got to the black hole. Eventually there would come a point, known as the static limit, at which the space traveller would have to move faster than the speed of light with respect to the black hole in order to remain stationary with respect to the rest of the universe. Inside the static limit, therefore, the space traveller must be carried around by the black hole.

The static limit is outside the event horizon, and so another way in which black holes might lose energy has been suggested. In the Penrose process, an object approaching a black hole might cross the static limit, and then split into two, with one half going into the black hole and the other re-emerging with more energy than the original whole particle. The increased energy of the emerging particle would have been obtained by decreasing the rotational energy of the black hole. The strict limitations, however, on the behaviour of the particles, if energy is to be extracted in this manner, make it unlikely to be a significant source of energy in practice.

A more straightforward and probable energy-generating mechanism is that of the accretion of matter into a black hole. Such material would generally initially form an accretion disc orbiting the black hole (like that around the white dwarf in nova systems, Journey 6). Eventually the material would fall into the black hole as viscous drag caused the orbits to decay. But prior to that, the potential energy released during the fall would go into heating the accretion disc, and thus become available to power astrophysical processes. Since up to 42% of the rest mass energy $(3.8 \times 10^{16}$ J from each kilogram, or 60 times the energy obtained by converting 1 kg of hydrogen to helium) of the accreting material may be released in this way, it may provide a mechanism to explain the energy emissions of very luminous compact sources (Journeys 9 and 10).

After all this theoretical speculation we finally turn to the question of whether we can find any 'real' black holes. As mentioned at the beginning of this journey, actual examples of black holes have yet to be found unequivocally. There are, however, several objects wherein the presence of a black hole is strongly suspected. The main line of reasoning is based upon the mass limits for compact objects. Electron degeneracy pressure is unable to support a white dwarf against gravity once its mass exceeds the Chandrasekhar limit of about 1.4 solar masses (Journey 6). Neutron stars similarly have an upper limit to their masses, and this is variously estimated to lie between two and three solar masses (Journey 7). A neutron star of over three solar masses must therefore eventually collapse to a black hole, and any compact object with a mass greater than this must be a strong black hole candidate.

The most promising object of this type is the binary star HDE 226868 (figure 8.4), also known as Cyg X-1 from its x-ray emission. The optical primary is a hot supergiant with a mass of about 20 solar masses. The most probable mass for its companion is about 12 solar masses with an almost absolute lower limit of 3.3 times the mass of the Sun. This companion must also be a compact object since its variations on a timescale of milliseconds† place an upper limit on the size of the emission region of 1 000 km. Although alternative explanations for the observations have been suggested, including the presence of a third body

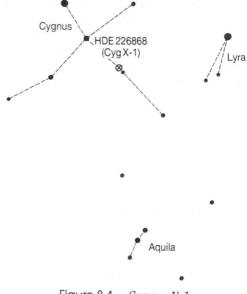

Figure 8.4   Cygnus X-1.

---

† In general, the observed changes in the brightness of an object must take a finite time to occur, the minimum time for a significant change being the time for light to cross the object (Journey 10). Thus if an object changes its brightness by a significant amount in a millisecond, its size must be less than 300 km $(0.001\,\text{s} \times 300\,000\,\text{km}\,\text{s}^{-1})$.

in the system, these require very severe constraints upon the parameters of the system, and the companion of HDE 226868 is probably best interpreted as a black hole.

V861 Sco is a very similar system to HDE 226868. It has another hot supergiant in a binary system with a period of about eight days. The companion again seems to be a compact object, this time with a mass about eight times that of the Sun. The x-ray transient A0620-00 may have a seven solar mass black hole with a faint low mass companion. Other less clear-cut cases include Cir X-1 and LMC X-3. Finally, some of the models for the intriguing emission line object SS433 have a 0.4 solar mass white dwarf in orbit around and being pulled apart by a five solar mass black hole (figure 5.12).

Rather less clearly, there may be evidence for black holes with masses millions of times that of the Sun. Jets of material are found being emitted from many galaxies, and may originate from massive black holes. The centre of the Seyfert galaxy NGC 4151 seems to have a mass of 10^9 solar masses in a region about 3×10^9 km across, and this is just about the Schwarzchild radius for such a mass. Many of the theories purporting to explain the energy emissions from quasars, Seyfert galaxies and even from the centre of the Milky Way galaxy (Journeys 9 and 10), require massive black holes in order to produce the observed energy emissions from the observed volumes.

Finally, as mentioned earlier, 'mini' black holes may be disappearing with an intense burst of gamma radiation if their initial masses were around 10^{11} kg at the time of the big bang. However, though many bursts of gamma rays have now been detected by spacecraft-borne detectors, none has yet had the characteristics to be expected from a dying black hole. Thus, sadly, at the time of writing, the observational case for the existence of black holes remains 'not quite proven'.

Journey 9

To the Hub!

> *If then, Socrates, in many respects concerning many things—the gods and the generation of the universe—we prove unable to render an account at all points consistent with itself and exact, you must not be surprised. If we can furnish accounts no less likely than any other, we must be content.*
>
> Timaeus
> Plato

Our last journey took us well beyond the edges of conventional science, and into realms which must have seemed at times like extracts from some of the more outré science fiction novels. It may come as something of a relief therefore to return to 'real life', and set off on our next journey, towards the centre of our own galaxy.

Such relief though, will prove short-lived. For what should be waiting for us at the end of our journey, at the hub of that enormous Catherine wheel, part of which we see as the Milky Way, but yet another black hole. To see why we believe there is a black hole at the centre of the galaxy, for like all other black holes its existence remains a plausible inference, not a proven result, we must start in the more conventional environment of the space around the solar system.

On any clear night, especially if there is no moon and you are away from artificial lights, an irregular luminous band can be seen stretching from horizon to horizon, faint but quite clear. This is the Milky Way. One of Galileo's first observations with his newly invented telescope was to show that this amorphous band of light was actually composed of millions of faint stars. Thus in an instant, and perhaps regrettably, the stream of milk spilling from Cassiopeia's breast (figure 9.1) became myth. Though too dim to be seen as individuals with the naked eye, the combined effect of these faint stars has nonetheless continued to inspire poets and lovers. The ancient myth has been retained in our name Milky Way, and even in the term Galaxy which is derived from the Greek word for milk: γαλα.

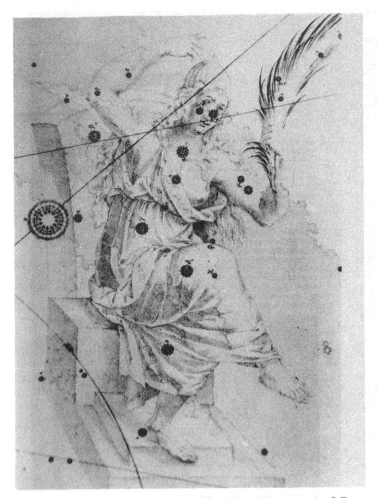

Figure 9.1   Cassiopeia and the Milky Way (*Uranometria*, J Bayer, 1603).

On any given night only a portion of the Milky Way may be seen. Over several months though, most of it may be photographed or drawn. From the United Kingdom, the whole of it can never be observed, the southern-most portion always remaining below the horizon. Observers nearer the equator, between latitudes of about 25° N and 25° S, can observe the Milky Way in its entirety (figure 9.2). We find then that it is very irregular, with light and dark patches intermixed throughout. It is also noticeable, however, that overall the Milky Way brightens towards the south, peaking around the southern constellation of Scorpius. As we shall see, the actual centre of the galaxy occurs in Sagittarius, close to its

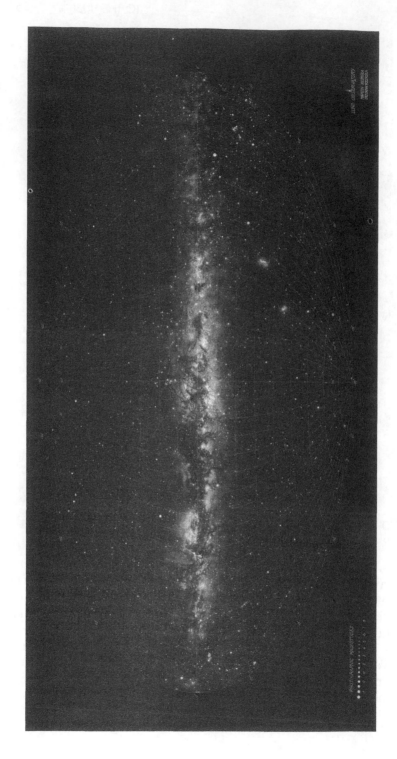

Figure 9.2   Panorama of the Milky Way. (Reproduced by permission of the Lund Observatory, Sweden.)

border with Ophiuchus (figure 9.3), and is visible to UK observers low in the southern sky on summer evenings soon after sunset.

Once Galileo had shown that the Milk Way was a vast aggregation of stars, it became natural to enquire about its three-dimensional shape, and the whereabouts of our solar system within it. William Herschel was the first astronomer to make a serious attempt at answering this question. His method of 'star-gauging' assumed that stars were uniformly distributed in space. If more stars could be observed in one direction than in another, then it must be, on this hypothesis, that the Milky Way collection of stars extended further out into space in that first direction than in the second.

Herschel's assumption is of course incorrect; it is clear that sometimes more stars are seen in some directions than in others because we see a closely packed cluster of stars (figures 3.16 and 9.4). Herschel, however, was aware of the problem, and in selecting the 1 100 or so areas of the sky in which to count stars he avoided the obvious clusters, and he averaged the results from several nearby areas in order to reduce the effects of the unobvious clusters.

Herschel found that the star system was roughly shaped like a long thin flat box split towards one end (figure 9.5). He furthermore found that the solar system was close to its centre. Now ever since Copernicus

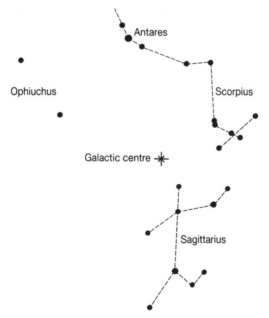

Figure 9.3   Finder chart for the position of the galactic centre in the sky.

Figure 9.4   The globular cluster M13 in Hercules: a close-packed aggregation of about a million stars. (Official US Naval Observatory photograph.)

and Kepler overthrew Ptolemy's geocentric hypothesis, astronomers have regarded with suspicion any result which tended to place mankind at some special point in the universe, labelling such ideas anthropocentric. But despite the doubts thus caused by the Sun's apparent proximity to the centre of the star system, Herschel's picture remained largely unchanged for some 130 years.

Something closer to the present-day view of our galaxy started to emerge only in the second decade of this century. Harlow Shapley pointed out that the globular clusters (figure 9.4) were to be found predominantly in the southern part of the sky, a third of them occurring within just the constellations of Sagittarius and Ophiuchus. He suggested that they formed a spherical halo centred on the core of the Milky Way galaxy. By finding the centre of the halo he went on to show that the Sun, far from being close to the middle of the galaxy, must actually lie way out towards its edge. Indeed, we now place ourselves some 26 000 light years from the centre of the galaxy: almost halfway to the outer edge of its disc.

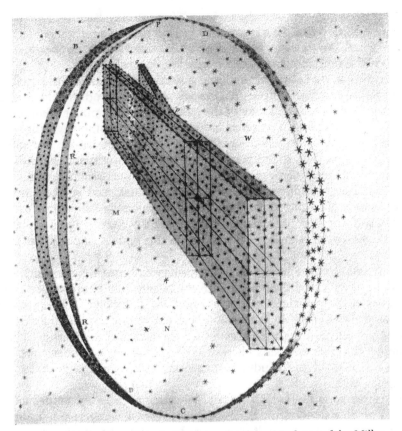

Figure 9.5   William Herschel's estimate of the 3D shape of the Milky Way star system, based upon his 'star gauging'. The dot near the centre shows the Sun's position. (1784 *Phil. Trans.* **74** 437.)

Why did Herschel and other observers up to Shapley's time get the position of the solar system inside the Milky Way so wrong? The answer, as always, is obvious once known.

The dark regions to be seen in figure 9.2 do not arise from the absence of stars in those particular directions, but from the obscuration of the more distant stars by huge clouds of dust particles. Such dust clouds permeate most of the galaxy, and even the brightest portions of the Milky Way have their more distant stars hidden.

In the optical region of the spectrum, therefore, we can only see a small section of the whole galaxy. Herschel's observations were limited to stars within a radius of about 7 000 ly of the Sun, and his resulting model for the star system was thus naturally centred close to the Sun, representing only that portion of the galaxy near the Sun: about 1% of the whole.

In the last half-century, the features of our galaxy have become known in much more detail. This is not because our optical observations are able to penetrate the dust any better than in Herschel's time, but because new areas of the spectrum have been opened up to observation.

Photons in the radio, infrared and gamma-ray regions of the spectrum are able to pass through the dust clouds with little hindrance, and so can be used to map out the whole of the galaxy. The first major advance over the optical work occurred when the radio emission at a wavelength near 21 cm was detected. The significance of this emission (actually at 21.106 cm) is that it originates from neutral hydrogen. Though cold hydrogen fills the whole of interstellar space, its emissions are naturally brighter from the denser clouds of gas and dust, and so these may be plotted out. The concentrations of stars probably follow the concentrations of the hydrogen clouds. Hence such a radio map of the galaxy (figure 9.6) will probably be similar to the optical appearance of the galaxy.

It should be noted that parts of the galaxy in figure 9.6 are omitted, not because they cannot be observed at 21 cm, but because the orbital motions of the hydrogen clouds around the galaxy in these regions are across the line of sight. The emissions from clouds at differing distances from the Sun cannot then be separated from each other by their differing Doppler shifts as happens elsewhere in the galaxy. Furthermore, individual motions of the clouds cause considerable blurring and uncertainty in the picture.

Infrared observations, however, have more recently confirmed and added further detail to the radio picture. Thus if we were able to look at our galaxy from outside, it would probably have an appearance similar to M81 (figure 2.2(a)).

Having thus got our map to hand, we can finally set out on our journey towards the hub of the galaxy. But let us not be in too much of a hurry. Let us go along the 'pretty' route.

We thus start by moving off towards the faint constellation of Coma Berenices, mid-way between the better known constellation of Leo and the star Arcturus (in Bootes). This will take us out of the galaxy on a line perpendicular to the disc of the spiral arms. We soon find that the stars are becoming few and far between. After only 700 ly the density will have dropped from one star in 100 cubic light years to one in 400. While at 3 000 ly, only 12% of the distance from the Sun to the centre of the galaxy, the density will have fallen to 3% of its value near the Sun. By the time that we have reached a distance at which we can comfortably see the Milky Way galaxy appearing like M81, say 300 000 ly out from the disc, a third of a million cubic light years will need to be searched to find each star.

At this point, 300 000 ly out, the Milky Way will be just visible to the

Figure 9.6   21 cm map of the Milky Way galaxy. The position of the Sun is shown by the encircled dot. (Reproduced by permission of Prof. G Westerhout, US Naval Observatory.)

naked eye as a faint spiral glow about 20° across: about the size of the main portion of the constellation of Orion as seen from Earth. Additionally we should be able to see several small galaxies, such as the Magellanic clouds (figure 7.1), as satellites of the main galaxy at distances up to 750 000 ly out. Closer in, there are also several hundred globular clusters (figures 3.16 and 9.4).

These surrounding small galaxies and clusters are the main signs that

Figure 9.7   A spiral galaxy (NGC 4565) viewed edge-on. (Reproduced by permission of the Palomar Observatory.)

the Milky Way galaxy extends far outside its normally visible limits (cf figure 9.7). Only in the last few years has it become apparent that the visible portion of the galaxy is probably only 10% of the whole. The outermost portion, through which we are still travelling on our journey, is called the Corona. This extends at least 300 000 ly from the galaxy, and may contain as much as 10^{12} times the mass of the Sun, or about half the total mass of the Milky Way system. Some of this material is in the form of the small satellite galaxies, but most has yet to be directly detected and may be in the form of old, cool, faint, burned-out stars.

The centre of the corona increases sufficiently in density to be classed as a separate region: the halo. This contains up to 10^{11} solar masses in a slightly flattened sphere about 100 000 ly across. About half of the globular clusters are found in the halo, along with isolated old stars. Without the presence of this massive halo, the disc and spiral arms of the Milky Way would be gravitationally unstable. We must thus expect all spiral galaxies to have similar halos and perhaps coronae, and in some cases these have been detected (figure 9.8).

Having taken this detour in order to view the Milky Way in its entirety, let us now head back towards its centre. At 10 000 ly from the centre, well inside the central bulge of the galaxy, we encounter the first sign that all may not be peaceful when we arrive. At this point we travel through an incomplete ring of neutral hydrogen. The material is rotating

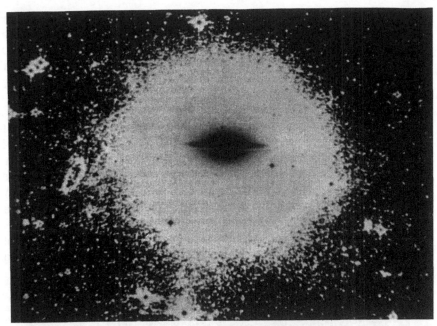

Figure 9.8 The halo around the Sombrero galaxy: a negative image of the galaxy has been superimposed upon a positive image of the halo. (Reproduced by permission of the European Southern Observatory.)

about the centre of the galaxy and expanding outwards at about 90 km s^{-1}. Perhaps it is the start of the formation of a new spiral arm, perhaps the result of an explosion in the centre of the galaxy a few tens of millions of years ago, perhaps even both. Alternatively, the ring may be due to some other, as yet entirely unknown, process.

Further on in, 5 000 ly from the centre, we come across another expanding rotating disc of cold atomic and molecular hydrogen. This disc is tilted at 20° to the plane of the galaxy. One thousand light years out is yet another ring. But parts of this latter ring have been heated by the presence of young supergiant stars embedded in it until its hydrogen has been ionised. Finally at 30 ly from the centre is the last ring, again formed of ionised hydrogen, but with only about half the temperature of the 1 000 ly ring: 5 000 K.

Of course, all this while we have also been passing the 'normal' constituents of the galactic bulge: a dense aggregation of old, small stars and a thin interstellar medium of gas and dust. Additionally, large, dense clumps of cold hydrogen molecules and dust known as giant molecular complexes may be found, mostly towards the outer edge of the bulge 16 000 ly from the centre.

We can now see in front of us the central 10 ly of the galaxy. From our vantage point 30 ly out, it totally dominates our view. More than a million stars, each many times brighter than the Sun, are crammed into a sphere, a star density a million times that near the Sun. Even this far away it provides us with an illumination as good as early evening twilight on the Earth.

If we switch to radio wavelengths then the stars fade and we see streamers of ionised gas (figure 9.9) falling inwards, from a turbulent ring of gas and dust about 6 ly in radius, through a vacuum that is hard even by interstellar standards. The streamers are moving at a few hundred kilometres per second and accelerating as they get nearer the centre. They are visible from the Earth as the radio source Sagittarius A West. At shorter wavelengths we find the scene illuminated by gamma rays produced as electrons and positrons annihilate each other.

The movements of the gas streamers suggest that right at the centre of the Milky Way galaxy is a very compact object with a mass at least a

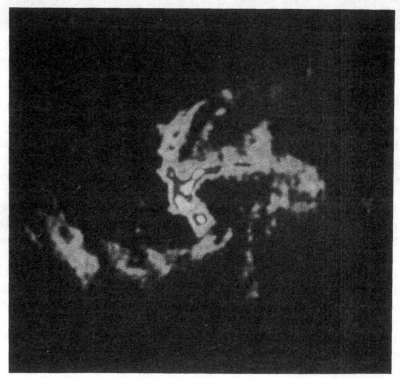

Figure 9.9   A radio pseudophotograph of the innermost 10 ly of the Milky Way galaxy. (Reproduced by permission from *Science* **320** 159 (1985) by M M Waldrop. Copyright AAAS.)

million times that of the Sun. In the radio region we know it as Sagittarius A* and we can see that it is less than 20 AU across: the size of Saturn's orbit.

Even now we cannot be sure that it is a black hole; collections of neutron stars or hot massive young stars might provide most of the observed phenomena, except for the total mass. But such a concentration of massive objects in such a small volume would almost certainly have collided, coalesced and formed a black hole long before now.

Thus a black hole is the most probable identity of the central object in the galaxy. The motions of the streamers then have an attractive possible explanation. Material falling towards the black hole might very well have caused an explosion sufficient to clear the 6 ly around the black hole of matter. The material from such an explosion 10 000 years ago could now just be starting to fall back towards the black hole. The observed streamers would be this infalling material and the explosion could have left the 6 ly ring in its observed turbulent and disturbed condition. Perhaps in due course the infalling material will produce another explosion and the whole process may prove to be cyclic.

Although a million solar mass black hole is a massive object by normal standards, it is only a millionth of the mass of the galaxy; and yet we find that it is very close indeed (a few light years) to the centre of rotation of the galaxy. It is inconceivable that the presence of the black hole should cause the galaxy to rotate around that point. Rather there must be a cause for the black hole to form at the centre of rotation of the galaxy. Just such a mechanism for this latter process is ready to hand in the coalescence of stars from the millions to be found in the central 10 light years. The stars closest to the centre have the lowest angular momenta, and are the ones most likely to collide, and so the black hole would come into existence close to the centre of rotation. Most probably the black hole formed very early on in the life of the galaxy, perhaps even before most of the stars condensed.

Now by the anthropocentric principle which we encountered earlier in this journey, we do not expect to find ourselves in any particularly special place in the universe. But if our own local galaxy has a massive black hole at its heart, what of other similar galaxies and of the plethora of more exotic objects? It is becoming increasingly clear that most, if not all, galaxies, QSOs, quasars, Seyferts, Liners etc are probably centred on black holes, and that in some cases these black holes may be thousands of times more massive than the one at the centre of the Milky Way galaxy. But therein lies another tale and its telling must await our next journey.

Journey 10

The Enigma Machines

. . . the German Enigma machine. This was a very ingenious arrangement of three wheels, each one of which had a sequence of studs on each side, with each stud on one side being connected by a wire to a pin on the other side—the exact arrangement of the connections being one of the secrets of the machine—and the pin making contact with one of the studs on the next wheel. . . . every time the machine was operated to encode a letter, one wheel would be turned by one space; after this wheel had . . . moved through one revolution, it would click its neighbouring wheel by one space. . . . the appropriate key would be pressed on the keyboard, and the resultant coded letter would be determined by the appropriate conducting path through the studs. . . . A further touch of ingenuity was to add a reversing arrangement at the edge of the third wheel . . . so as to send the current backwards through the wheels by yet another path. The returning current lit a small electric bulb which illuminated a particular letter on a second keyboard, and thus indicated the enciphered equivalent of the letter whose key had originally been pressed.

Most Secret War
R V Jones

Such a machine would surely have been a joy to Verne's Professor Liedenbrock and his Ruhmkorff's apparatus, but despite its complexity the Allies' cryptographers were able to decode messages produced by it from early on in the second world war. Today's enigma machines are less martial but far more puzzling. Their existence was unsuspected during the second world war, and indeed for another two decades afterwards. They are those distant powerhouses known variously as quasars, QSOs, Seyfert galaxies, BL Lac objects, active galactic nuclei (AGNs), optically violent variables (OVVs), and by many other titles. They are linked by the common property of producing prodigious amounts of energy from very small regions. In some cases energy at a level thousands of times that radiated by the whole of our Milky Way galaxy comes from a region comparable in size with the solar system.

The story starts in 1960, when the 48th object found during the third survey of the sky at radio wavelengths by the Mullard Radio Observatory of Cambridge, and thus known as 3C48, was identified in the optical region with a bluish star-like object (figure 2.3).

The spectrum of this blue 'star' turned out to be most mysterious: the lines in it were all emission lines (an unusual but not unknown situation for 'normal' stars). Rather more startling, however, was that none of those lines could be identified with lines produced by any of the known chemical elements.

The spectrum of 3C48 was to remain a mystery for three years until a second, similar, object was discovered. This was another radio source found by the third Cambridge survey: 3C273. The lines in its spectrum turned out to have been produced by hydrogen, but red-shifted (Journey 2) by what then seemed the enormous amount of 16% (figure 10.1). Given this clue, the spectrum of 3C48 could then be recognised as also being produced by hydrogen, but red-shifted by 37%. The first of the Balmer lines of hydrogen (Hα) was thus moved into the infrared and was no longer visible. The second line (Hβ) normally found in the yellow-green part of the spectrum was moved into the deep red, and so on.

Because of their star-like appearance and their radio emissions, 3C48 and 3C273 became the first members of a class of sources soon to be called **qua**si-**s**tellar **r**adio **s**ources, or quasars. In their turn quasars are just a part of a much larger group: quasi-stellar objects or QSOs. The QSOs

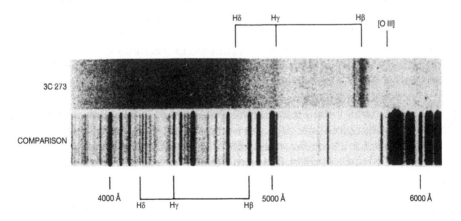

Figure 10.1   The spectrum of 3C273 (negative photograph). (Reproduced by permission of the Palomar Observatory.)

are star-like blue objects with highly red-shifted emission lines. The quasars are a subgroup of the QSOs with strong radio emission.

Now as we have seen previously, changes in the observed wavelengths of spectrum lines can arise through relative motion of the source along the line of sight (Doppler shift). The red-shifts of 3C48 and 3C273 require them to have velocities away from us of 30%† and 15% of the velocity of light (90 000 and 45 000 $km s^{-1}$) respectively. The current record holder, PC 1158+4635, has a red-shift of 4.73, corresponding to a velocity of 282 000 $km s^{-1}$.

The question then arises of whether these recessional velocities are due to the distances of the QSOs away from us and are therefore just a part of the general expansion of the universe (Journeys 1 and 2), or whether the QSOs are much closer to us and have acquired their enormous velocities by some other process. For a long while the answer to that question was unclear, but it is now almost certain that the red-shifts are cosmological and that the objects are indeed at enormous distances.

The importance of determining the distances of QSOs, or at least of deciding whether they are local objects or at cosmological distances, arises because evaluating all their other properties is dependent upon this one datum.

Thus if QSOs are local (i.e. within a few million light years of our galaxy), then their luminosities lie within the range possible for stars. The mystery then lies in finding out how they are accelerated to such high velocities and why they are all moving away from us, for none have yet been found with blue-shifts.

On the other hand, if QSOs are very distant (thousands of millions of light years), their velocity arises naturally from the expansion of the universe. The mystery then lies in their extreme brightnesses: in some cases thousands of times brighter than the total energy emission from a large galaxy like the Milky Way, and in this energy having to come from a very small region. An upper limit to the size of the emitting region is set by the time taken for significant variations in the luminosity of the object (Journey 9). For some QSOs this can be as short as a few hours, limiting their sizes to a few tens of astronomical units, or a volume in the region of 10^{-25} of that of a galaxy.

The evidence for the cosmological distances of QSOs arises on several fronts. For example, some QSOs are found accompanying distant clusters of galaxies and with similar red-shifts. The probability of such an association along the line of sight, but with the cluster and QSO at very different

---

† Not 37% and 16% of the velocity of light (c) because at very high velocities the simple formula for the Doppler shift becomes inaccurate, and a relativistic version is required. Thus at a red-shift of 100% ($z+1$) the velocity is $0.6c$, at $z=2$ it is $0.8c$, at $z=3$ it is $0.88c$, at $z=4$ it is $0.92c$, and so on.

distances from us, occurring by chance is small. Thus these QSOs almost certainly lie within their attendant clusters. Other QSOs are deduced to lie behind distant galaxies, either from the presence of absorption lines from the gas in the galaxy in the spectrum of the QSO, or through the production of multiple images of the QSO by the gravitational lens effect† of the galaxy (figure 2.13).

Many QSO optical spectra, then, contain tens or sometimes hundreds of narrow absorption lines. They are all due to absorption by hydrogen in its lowest energy level which results in the Lyman-alpha line. This line is normally found far into the ultraviolet at 121.5 nm (few people can see wavelengths shorter than about 380 nm). The lines, sometimes called the Lyman-alpha forest because they are so numerous and so close together, each then arise by absorption in a cloud of cool hydrogen gas between us and the QSO. Individual clouds are moving away from us at different velocities from other clouds, and so their Lyman-alpha absorptions are red-shifted to different degrees. Since the velocities of the clouds are large with respect to the QSO, it is difficult, given their low temperatures, to imagine that they have been flung out from it. They must therefore just be intergalactic clouds whose velocities result from the general expansion of the universe. The QSO lying beyond them must therefore also be very distant.

A final argument is less direct, but is nonetheless very strong. It is that the known quasars and radio galaxies together already provide almost a third of the total radio energy reaching us from space. To avoid such discrete sources from eventually adding up to more than the total radiation we receive, we must already be seeing them out to distances nearly comparable with the distance to the edge of the visible universe.

Thus we have an initial picture of QSOs as objects occupying a

---

† As we saw on Journeys 2 and 8, light and other electromagnetic waves have their paths affected by gravitational fields in a similar manner to those of material particles. If for an observer, then, there are two objects along the same line of sight, the light paths from the more distant one will be bent by the gravitational field of the nearer. Unlike a normal optical lens, however, the bending of the light paths reduces with increasing distance away from the optical axis, and so a true image is not produced. The optical equivalent of the gravitational lens is given by the stem and base of a wine glass (figure 10.2). If the two objects are exactly on the same line of sight, then the more distant would appear as a uniform ring surrounding the nearer. If, however, the alignment is not quite perfect, then the image of the more distant object will be asymmetrical and the ring will break up into two or three short sectors. These will not normally be resolvable and so the more distant object will then appear as two or three point sources close in the sky to the nearer object (though the latter often may not be visible).

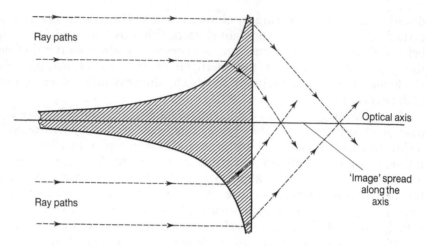

Figure 10.2   An optical equivalent of a gravitational lens.

relatively small volume, a few tens of astronomical units to a few light years across, but emitting up to 10^{41} W. Their distances range from a thousand million light years to at least ten times that figure.

However, that is not quite the whole story. Many QSOs, upon close examination, turn out to have resolvable structure at both radio and optical wavelengths. Thus 3C273 has a jet emerging from it (figure 10.3). Other QSOs, including 3C48, appear to be embedded within 'fuzzy' nebulosities which sometimes turn out to be fairly normal galaxies. Often at radio wavelengths additional emission is found coming from two large regions on opposite sides of the QSO. They are thus not truly 'quasi-stellar' (i.e. nearly point sources), but sometimes have associated structures extending outwards by tens or hundreds of thousands of light years.

Furthermore, QSOs are not the only types of object to be emitting large amounts of energy from small volumes. As we saw on the last journey, there is evidence of intense activity, maybe even an explosion, having occurred within a small volume at the centre of our Milky Way galaxy in the relatively recent past.

More importantly, there are other types of galaxies which seem to possess the properties of QSOs in reduced form. Thus the Seyfert galaxies (figure 2.3) have cores to their nuclei which may be under a light year in size and yet emit up to a hundred times the total amount of energy coming from the rest of the galaxy. Some of the Seyfert galaxies are also strong radio emitters, and the spectra of their cores reveal strong, broad emission lines like those of the QSOs.

Then we have the BL Lac objects which may be even more luminous than QSOs. Their optical spectra are almost featureless continua, while in

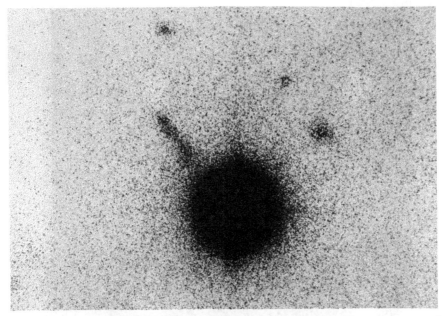

Figure 10.3   QSO 3C273 (negative photograph). (Reproduced by permission of the Palomar Observatory.)

the X-ray and radio regions their emissions vary wildly on timescales ranging from a few tens of seconds to a few days.

By no means finally, but the last objects to be considered here, are the radio galaxies (figures 1.7 and 2.4). These differ from the other objects just mentioned in not seeming to have compact energy sources, but they do often have double emission lobes and jets somewhat similar to those of the QSOs.

One of the most surprising observations in recent years has been that the jets associated with QSOs seemed to be moving away from their central objects faster than the speed of light, a phenomenon called superluminal expansion. In the case of 3C273 (figure 10.4), the jet seems to have moved 25 light years in only three years of time: over eight times the speed of light.

Fortunately for special relativity, the light speed barrier is not actually being broken. The explanation of the apparent superluminal velocity is that a jet of material has been expelled from the central object almost directly towards us at close to, but at less than, the speed of light. The material in the jet is therefore only lagging behind the radiation that it emits by a small fraction of the speed of light. The radiation emitted later in time thus comes from a point much closer to us than that emitted

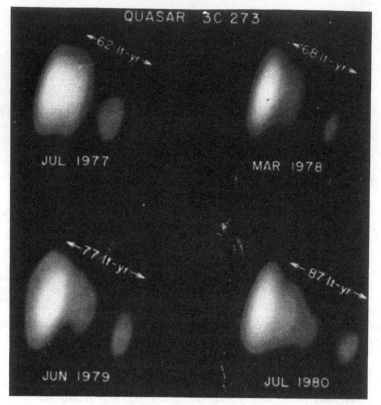

Figure 10.4   Superluminal expansion of the jet in 3C273. (Reprinted by permission of *Nature* **290** 365, T J Pearson *et al*, Copyright 1981 Macmillan Journals Ltd. Courtesy of California Institute of Technology.)

earlier, and so it arrives over a much shorter time span than that over which it was emitted (figure 10.5).

Until recently, the various disparate denizens of the outer reaches of the universe appeared to be only marginally inter-related. Now, however, it may be that they are really all the same or very similar types of object, but viewed from different directions. The link is suggested by the jets that are associated with many of them.

Let us leave aside for the moment the question of the source(s) of the energy of these objects, and just assume that it is possible to produce the required immense amounts of energy from the required small volumes. Given such a powerhouse, can the rest of the phenomena be explained?

A recent suggestion by Peter Barthel suggests that a single model can indeed be found which fits the observations of QSOs, BL Lacs, Seyfert galaxies, radio galaxies and maybe even of spiral galaxies like the Milky

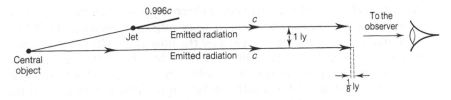

Figure 10.5   Superluminal velocities: one configuration to give an apparent velocity eight times the velocity of light.

Way. A theory with such a powerful unifying theme is not automatically correct, but it does seem likely to be on the right lines. Barthel's model (figure 10.6) has the central powerhouse surrounded by a thick ring of dense dust clouds some hundreds of light years across. Two intense jets of relativistic electrons, and possibly other particles as well, are ejected from the powerhouse along the central axis of the dust cloud torus. This whole structure is then embedded within a relatively normal, though sometimes turbulent, spiral or elliptical galaxy.

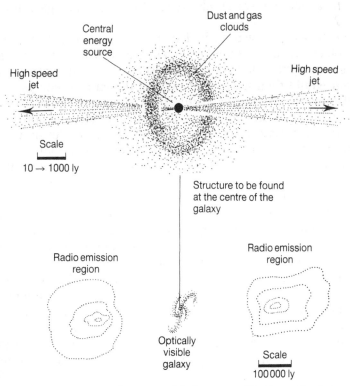

Figure 10.6   Barthel's unifying model for QSOs, BL Lacs, Seyfert and radio galaxies.

A Barthel object will present very different appearances when viewed from different directions. Observed at right angles to its axis, the bright central powerhouse will be obscured by the dust clouds in the surrounding ring. The jets, however, will emerge visibly on both sides, though they will appear faint because the radiation from particles moving at close to the speed of light is mostly emitted into narrow beams along the directions of motion. The electrons in the beams may eventually be brought to a halt as they collide with denser clouds of gas, depositing their energy into the gas and producing strong radio emission. Such gas clouds could have resulted from an earlier explosion in/around the central powerhouse. The torus of dust clouds would prevent the expansion of the gas in the 'equatorial' regions, leaving it to emerge just along the axial directions. An alternative explanation could have the velocity of the jets increasing with time, the currently observed relativistic particles then being effectively brought to a halt as they catch up with the more slowly moving material emitted earlier (a similar effect may result in the ring types of planetary nebula, Journey 5). Either way, a pair of symmetrically placed diffuse radio sources will be formed on opposite sides of the central object, just as is observed with many radio galaxies (figures 1.7 and 2.4).

If, by contrast, we were to be looking along or near to the axis of a Barthel object, then we would be looking directly into the emerging jet. The beaming of the radiation by relativistic particles along their direction of motion would then mean that most of the emitted energy would be directed towards us. The jet would thus appear extremely bright and would swamp the other emissions, in particular the emission lines from the gas and the radio emissions from the lobes. The jet moving away from us would, of course, not be visible because its radiation would be being beamed directly away from us. With the line of sight along the axial direction therefore, we would observe the system to have the appearance of a BL Lac object.

Finally, what if the viewing angle lies between the axial and equatorial directions? Then, as may perhaps have been guessed by now, we would see the object as a QSO. Emission from the approaching jet would be bright, but we would not be close enough to the main beam for it to be dominant. Thus we would be able to see both the bright core around the central powerhouse and the emission lines from the surrounding gas, as observed in typical QSO spectra. The two radio lobes would no longer be superimposed along the line of sight and swamped by the jet's emissions as in the BL Lac objects, but would be seen on either side of the QSO. However, only the approaching jet would be visible. The receding jet, which produces the lobe on the far side of the QSO, would be too faint to detect because its radiation would be directed away from us (though some QSOs with genuine one-sided jets also seem to exist). The radiation from the more distant lobe would have its emission 'scrambled' by having to pass through the hot gas around the QSO—a prediction which fits in very well with the otherwise puzzling observation that the radiation from one of the radio lobes of QSOs is usually much more strongly polarised than that from the other.

Thus in a single model we are able to explain many of the properties of apparently quite dissimilar objects. There is also some further supporting evidence for the validity of the model.

Firstly, since the model suggests that we are looking at radio galaxies side-on, but at QSOs at a narrower angle to the axis, the average separation of the radio lobes should be greater for the radio galaxies than for the QSOs, and this indeed is exactly what is observed.

Secondly, the ratio between the number of radio galaxies and QSOs suggests that the viewing angle for QSOs should be less than $45°$ to the axis of the system. Thus with QSOs we should always be looking into the relativistic approaching jets at fairly narrow angles to their directions of motion. This is exactly the situation required to produce superluminal expansion (figure 10.5) and, just as the model predicts, of the QSOs so far observed in sufficient detail to provide accurate data, all have super-luminal motions.

Barthel's model, however, is not without deficiencies, most notably in that it provides no obvious reason as to why a few QSOs should be strong radio emitters from the central object (quasars), while most are relatively radio-quiet.

The relevance of the Barthel model to Seyfert galaxies is less clear. However, since the cores of the nuclei of such galaxies appear obser-vationally to be mini-QSOs, it may be reasonable to explain them via a scaled-down version of the model. Similarly, spiral galaxies in general, and the Milky Way in particular, which may undergo 'Seyfert-type' behaviour for some parts of their lives (Journey 9), could perhaps also have small versions of Barthel-type objects near their centres.

Now, successful though the model appears to be, it is dependent upon the initial assumption that a sufficiently compact powerhouse can exist, and that it can also produce the required relativistic jets of material. Fortunately, soon after the discovery of QSOs Donald Lynden-Bell proposed a mechanism for providing their energy. This mechanism, which is now widely accepted and which does have the required properties, operates via the infall of matter into a black hole which has a mass in the region of thousands of millions of solar masses.

We have already seen the evidence for a million solar mass black hole at the centre of the Milky Way galaxy (Journey 9), and motions of the gas clouds near to the centre of the Seyfert galaxy NGC 4151 suggest that it contains a mass of 5×10^8 solar masses within a volume 10^9 km in radius: roughly the size of the Schwarzschild radius for a black hole of that mass (Journey 8). Thus there is some direct evidence for the existence of massive black holes at the centres of some galaxies.

To produce the energies required for QSOs and similar objects, matter must be falling into the black hole (figure 10.7). As we have seen in other analogous but smaller systems such as the novae (Journey 6), the infalling material will not go directly into the central object, but will first form an accretion disc around it. The jets could then emerge as a result of gas and radiation pressure in the superheated central portions of the accretion disc. Expansion in the equatorial plane would be prevented by the material in the accretion disc, allowing escape only in opposite directions along the rotation axis of the black hole.

Energy can be emitted from the accretion disc itself, which will be at a high temperature, and also from additional matter falling inwards and colliding with the disc. Since nearly half the mass of material falling into a black hole may be converted into energy and radiated away, a bright QSO would require a few tens of solar masses falling into its central black hole each year to provide all its observed energy emissions. Such an accretion rate is enormous by human standards, but small compared with the typical mass of a large galaxy of 10^{11} to 10^{12} solar masses. Nonetheless, such a rate of destruction of matter is sufficient for us to

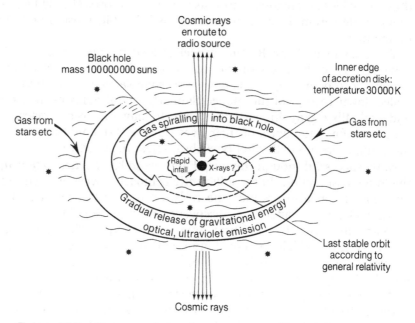

Figure 10.7   Energy emission from a massive black hole. (Reproduced from a SERC report, 1985, by permission of A Boksenberg and M Penston (RGO).)

expect that in the lifetime of the universe at least the central portions of the galaxy will have been swept clear. The infall of matter to the black hole should then reduce and we would expect the QSO activity to diminish or even disappear. Thus if QSOs were formed within a few aeons of the big bang, then we would no longer expect to see them nowadays.

Despite seeming appearances, such a decrease in QSO activity with time is exactly what is observed. Since the nearest quasar is almost 2×10^9 ly away, we are seeing it as it was two aeons ago, not as it is today. Most QSOs are five to ten times further away again, and so we see them as they were ten or more aeons ago. Only the less-spectacular Seyfert activity, or the homely little million solar mass black holes at the centre of the Milky Way and other spiral galaxies, are sufficiently frugal in their consumption of matter for their activities to have been occurring within the last few hundred million years while life has occupied the Earth.

Thus it seems probable that a detective story extending over nearly three decades is finally drawing to its close. With the Barthel model, the enigma machines may finally have been decoded.

Journey 11

She was the Universe

Quae visa, vera; quae non, veriora
(What you see is true; what you do not see is even more true)
                                                        Coromandel!
                                                         J Masters

During our various journeys, we have travelled backward to explore the rolling turmoil of the initial stages of the big bang, and forward to the cold and bleak ending of the solar system. Between those extremes we have taken in the formation of stars and galaxies, looked at some of the many ways in which stars come to an end, and plumbed the ultimate depths of black holes.

Nowhere have we yet looked at the fate of the universe as a whole; however if we take the scenario explored on the first journey of the hot big bang as the origin for the universe to be correct (and there are still some cosmologists arguing for alternatives to it) then we can project forward in time to see what may develop. There are two main possibilities: either the universe may continue expanding forever, or the expansion observed at present may some day come to a halt, and be succeeded by a collapse, perhaps back down to a singularity again.

Which of those possibilities lies in our future is determined by the amount of material in the universe (Journey 1). If the average density of the universe is greater than a critical density of about $10^{-26}\,\mathrm{kg\,m^{-3}}$ (equivalent to a few hydrogen atoms per cubic metre), then the gravitational field will be strong enough to turn the expansion into a collapse. If the average density is less than that critical figure, then the expansion of the universe will continue forever.

Now the average density of the universe is a quantity that we might hope to measure. In essence it would involve counting up the masses to be found in a volume sufficiently large to be representative of the universe as a whole, and then (mathematically) spreading that total mass evenly throughout that volume. The ratio of this observed mean density to the critical value is called the density parameter, Ω. In terms of the density parameter then, the universe will expand forever (an open universe) if Ω is less than one, and come to a halt and collapse (a closed universe) if Ω is greater than one.

Before looking at the observed values of Ω, let us return momentarily to the first journey, and remember how it was suggested that the universe could have originated as a quantum fluctuation of zero net energy. Should that be true, then it would mean that Ω was exactly equal to one, and the mean density of the universe was equal to its critical density. In that situation, the universe would continue to expand, with the rate of expansion slowing, so that after an infinite length of time the expansion would come to a halt. In practical terms, such an eventuality would be little different from the continually expanding universe obtained if Ω is less than one.

Determining the value of Ω from observations would seem to be a straightforward, if tedious, task just involving counting the galaxies over a volume of space at least a few tens of millions of light years across. Exactly such a process has indeed been attempted several times. In each case the mean density has turned out to be much less than the critical density, with values of 0.01 to 0.1 typically being obtained.

Thus at first sight, it would seem clear-cut that we are living in a universe destined to expand forever.

However, as we saw during Journey 2, observations of clusters of galaxies suggest that they contain far more material in some as-yet unobserved form than the total amount of material that we can detect. Similarly, on Journey 9 we saw that the Milky Way, and perhaps other galaxies as well, has a huge, faint halo containing far more mass than the visible parts of the galaxy. This 'missing mass' or 'dark matter' may amount to as much as 50 times the material making up the bright galaxies.

Suggestions for the form taken by the dark matter are many and various, ranging from heavy neutrinos, through other more massive subatomic particles, via large planets or small stars to million solar mass black holes and small galaxies.

Whatever the nature of the dark matter, if it exists in the postulated amounts then the value of Ω would become quite close to unity. Indeed, recent surveys extending out to several thousand million light years have suggested values of Ω as high as 0.9 with uncertainties of about ± 0.5. Such measurements are entirely consistent with Ω having an actual value of one, or even of having a value greater than one. Also, as we saw on our first journey, the unlikely coincidence of having the observed value of Ω close to unity has led some cosmologists to speculate that its true value will turn out to be exactly one.

Thus, at the time of writing, the best observational estimates of the value of Ω continue to allow all the possibilities: that it is greater than, equal to, or less than unity, and so cannot tell us whether we inhabit an open or a closed universe. Despite this ambiguity, we can, of course, still predict the possible futures resulting from any particular value of Ω; it is just not clear at the moment which future scenario is the one applicable to our universe.

Let us start then by looking at the future of a closed universe. That is to say, a universe which has a value of Ω greater than one.

In a closed universe the gravitational field of the matter and radiation contained within that universe will be strong enough for the expansion resulting from the big bang to be brought to a halt within a finite time. Thereafter the universe will collapse back towards its initial singularity. The length of time required for the expansion to be halted, however, is dependent upon Ω. The currently observed age of the universe of 10 to 20 aeons itself puts an upper limit on the value of Ω of 10. But as we have seen, the actual value is likely to be much closer to one. It will therefore be a very considerable time indeed before our currently expanding universe comes to a halt. It may well be that many of the events discussed for the future of an open universe will also have time to occur in a long-lived closed universe.

For the moment though, let us assume that the collapse will occur while at least some stars and galaxies remain in recognisable forms. The obvious question then is whether the collapse will simply be a reversal of the expansion. The answer to this appears to be no. Clearly, stars which have converted hydrogen into other elements and radiated away the resulting energy cannot easily be persuaded to reverse that process. Less obviously, there does not seem to be any inverse for the inflationary period of the big bang (Journey 1). The collapsing phase of the universe is thus likely to be longer, perhaps very much longer, than the expanding phase.

Throughout most of the collapse, little will happen that differs markedly from events in the expanding phase. Stars will be born and die, galaxies will evolve, more and more black holes will form, and so on. The main differences will be that distant galaxies will have their spectra blue-shifted, and that the microwave background radiation will be increasing its temperature.

Of course if all this has taken sufficiently long, the galaxies may have dispersed, or collapsed down to black holes, or even matter itself evaporated into radiation as we shall see below. However, continuing for the moment to assume that the universe is reasonably recognisable during its collapse, then about an aeon before the final moment (sometimes called the big crunch!), the clusters of galaxies would start to merge together. Collisions between galaxies would become more and more frequent, disrupting and changing their shapes. Eventually, perhaps a hundred million years before the end, the galaxies would lose their identities as separate objects and merge into a huge conglomeration of individual stars, planets and black holes, each of which would continue to rush towards final dissolution.

Between a million and a thousand years or so before the end, stars and planets would start to evaporate as the background radiation, no longer predominantly in the microwave region, reaches temperatures of 10 000 K or more, while at the same time black holes would form and grow catastrophically. In the last few seconds, temperatures and densities would reach those found soon after the original big bang, and the final state would be a fireball akin to that which started things off, but more disordered because of everything that has occurred within the universe since its origin.

What then will happen? Arguments continue over two possible fates for the fireball. It may, of course, just pinch itself out and disappear back into nothingness in a reversal of the original quantum fluctuation (Journey 1). An alternative suggestion, however, is that when it has shrunk down to a size of some 10^{-40} m or so it may 'bounce', creating a new big bang, and starting the whole process off again. It will not, however, be a simple repetition, for the time taken for each cycle from big bang to big crunch may increase, and compactification (Journeys 1 and 2) of the dimensions may proceed differently between different cycles. Such a model though, gives us an alternative to the quantum fluctuation as an origin for the universe, because there could then have been an infinite number of such cycles before our own. Thus the question of the origin would then be effectively eliminated by removing it to an infinite time in the past.

A final curious thought with which to end this part of our last journey: in a closed universe, nothing, not even the radiation, escapes the ultimate pull of gravity: the basic definition of a black hole. Thus if our universe is closed then we are arguably living inside a black hole!

The alternative to the closed universe is either an open universe or one which has a value of Ω of exactly unity. Although for the latter case the expansion will halt, it will take an infinite time to do so, and we can effectively include it with the open universes. In all these cases then, the expansion of the universe will continue, though slowing down as time passes. The smaller the value of Ω, the less the rate of expansion will reduce. Some of the times given below for events to occur will thus be reduced if Ω is significantly less than one.

We have seen (Journeys 4, 5, 6, 7 and 8) that stars come to the ends of their lives in a variety of ways. The more massive stars may only last for a few million years, while stars like the Sun may live for ten or twelve aeons, and smaller stars may live for longer still. Nonetheless, even the smallest star will eventually run out of nuclear fuel and shrink down to a white dwarf, before finally cooling to a black dwarf. At 10^{14} years, ten thousand times the present age of the universe, we therefore expect all stellar activity based upon nucleosynthesis to have finished. The stars themselves though, now the cold remnants of white dwarfs, will still mostly be aggregated into galaxies.

A thousand times older still, 10^{17} to 10^{18} years, and we will see significant changes occurring within the galaxies. Over such a period we may expect most stars to have undergone many close passages with other stars though, even over such an immense length of time, there will have been few actual collisions.

During a near miss between two stars, any planets surviving to the latter stages of the star's life are likely to be stripped away to become independent bodies. There is also the possibility of one of the stars acquiring sufficient kinetic energy from the other during their interaction to escape from the galaxy while the other star goes into an orbit closer to the galactic centre.

In this way many stars and planets will 'evaporate' from the galaxies to wander intergalactic space alone. Since by then the average distance between two galaxies will be approaching a thousand times the distance to the edge of the currently observable universe, that will be loneliness indeed.

The stars remaining with the galaxy will become more and more tightly packed because those being lost take with them more than their 'fair share' of the angular momentum. If a massive black hole does not already exist at the centre of the galaxy (Journeys 9 and 10), then it is likely to form now from the coalescence of the central stars. There may therefore be brief intervals of renewed activity and energy emission from the centres of galaxies as stars fall into the clutches of the black holes.

By 10^{20} years, a hundred times longer again and ten thousand million times the present age of the universe, gravitational radiation (Journey 2) will have caused any stars remaining in galaxies to have fallen to their centres. So the universe will just be populated by the massive black hole remnants of galaxies, the independent cold remains of the stars and planets etc that escaped from the galaxies, and the photons and neutrinos left over from earlier processes and from the big bang itself.

Now we saw on Journey 7 that neutrinos from the LMC supernova were picked up by experimenters who were trying to detect the decay of protons. That such decay can occur is by no means clear theoretically, and it has yet to be found observationally. If it does occur, then the proton lifetime is variously predicted to lie in the region of 10^{30} to 10^{35} years. If magnetic monopoles exist (Journeys 1 and 2), then they can disrupt protons and reduce their lifetime to 'only' 10^{20} years or so.

Thus, if protons decay by either process, then by 10^{35} years, the universe will contain matter only in the form of neutrinos, positrons and electrons (the positron is one of the decay products of the proton). Plus the super massive black holes left over from the galaxies.

Positrons are the anti-particles of electrons, and under today's conditions a positron very rapidly encounters an electron and the two annihilate each other, producing a pair of gamma rays. By the time we are now considering, however, the average separation of the positrons and electrons will be hundreds of thousands of light years. Their chances of encountering and destroying each other are thus negligible even on the timescales being considered here.

If protons do not decay then the black dwarfs and cold neutron stars will continue to exist in the ever-expanding universe. Eventually, however, in perhaps $10^{1\,000}$ years, the white dwarfs will collapse to neutron stars. The neutron stars in turn may collapse to black holes, in the inconceivable time of

$$10^{100\,000\,000\,000\,000\,000\,000\,000\,000\,000\,000\,000\,000\,000\,000\,000\,000\,000\,000} \text{ years.}$$

By 10^{70} years the electrons and positrons will be separated on average by a distance comparable with the size of the currently observable universe: 10^{10} light years or more. The relative energies involved, however, will be so small that the particles will form bound pairs in orbit around one another: effectively they will become electron/positron 'atoms'. The orbits will decay, however, causing the electrons and positrons to spiral in towards each other. They will therefore eventually annihilate each other and release gamma rays.

Once a black hole has formed it can lose energy by its Hawking radiation, as we saw during Journey 8. When the energy density of the rest of the universe falls below the level of the emission from the black hole then it will have a net loss of energy and will start to evaporate. The massive black holes left by the galaxies will take about 10^{100} years to disappear in this fashion.

Thus by 10^{100} years all the black holes will have gone, and the universe will contain only photons, neutrinos, and any remaining electron–positron 'atoms'. These last will eventually disappear as just described leaving just the photons and neutrinos. Now though electron–positron annihilation results in two high-energy gamma rays, the vast majority of the photons and neutrinos will be of very low energy indeed. By 10^{30} years, for example, the background radiation will have cooled to a temperature of only 10^{-15} to 10^{-20} K. Truly the weak shall inherit the universe!

We started our first journey with a question about the origin of time. It is fitting therefore to finish the last journey with a question about the end of time.

Our basic concept of time is based upon our observations that, overall, things progress from a more ordered state to a less ordered state. Thus we are familiar with the cup shattering into pieces, but we do not encounter broken cups spontaneously mending themselves. Of course, the cup had to be made in the first place, but that process would have involved increasing disorder elsewhere far more than the increase in order involved in the cup's manufacture: waste material would be dumped, heat lost to the atmosphere from the kilns and so on.

The physicist's term for the disorderliness of a system is called its entropy, and one of the most fundamental laws of physics, the second law of thermodynamics, is that entropy in a closed system increases with time. In the far future of an open universe, when we are left with just the photons and neutrinos distributed at random, it is difficult to see how the system (i.e. the universe) could become more disordered. At the end therefore, time will cease to have any meaning, because a state of maximum disorder cannot increase its disorderliness. There will be no means of knowing that time is passing, or even of knowing in which direction it is going!

Let us conclude our journeys with another quotation from Byron's *Darkness*. Though it was written in 1816, it still seems true today, and is likely to be even more true in 10^{100} years:

> *The waves were dead; the tides were in their grave,*
> *The moon, their mistress, had expired before;*
> *The winds were wither'd in the stagnant air,*
> *And the clouds perish'd; Darkness had no need*
> *Of aid from them – She was the Universe.*

Bibliography

Readers interested in pursuing topics encountered here may find some of the following books useful. The list is not exhaustive, but should enable a start on a literature search to be made. The Journey(s) to which a book has most relevance is also indicated.

Book	Relevant Journey
L F Abbott and S Y Pi *Inflationary Cosmology* World Scientific, 1986	1
H Arp *Quasars, Redshifts and Controversies* Cambridge University Press, 1987	1, 2, 10
P W Atkins *The Creation* Freeman & Co, 1981	1
M V Berry *Principles of Cosmology and Gravitation* Adam Hilger, 1989	1, 2, 8
F Close *End* Simon Schuster, 1988	1, 2, 10, 11
M Cohen *In Darkness Born* Cambridge University Press	3
N Cohen *Gravity's Lens* Wiley, 1988	10
P C W Davies *Space and Time in the Modern Universe* Cambridge University Press, 1977	1, 11
P C W Davies *The Accidental Universe* Cambridge University Press, 1982	1, 2

G F R Ellis and R M Williams   1, 2
Flat and Curved Spacetimes
Clarendon, 1988

A C Fabian   1, 2, 3
Origins
Cambridge University Press, 1988

S M Fall and D Lynden-Bell   2, 9, 10
The Structure and Evolution of Normal Galaxies
Cambridge University Press, 1981

G W Gibbons, S W Hawking and S T C Siklos   1
The Very Early Universe
Cambridge University Press, 1983

S Gibilisco   8, 9, 10
Black Holes, Quasars and other Mysteries of the Universe
TAB Books, 1984

J Gribbin   3
Galaxy Formation
Macmillan, 1976

J Gribbin   3, 4
The Strangest Star
Fontana, 1980

J Gribbin   1
In Search of the Big Bang
Corgi, 1986

S W Hawking   1, 2, 10, 11
A Brief History of Time
Bantam Press, 1988

B W Jones   3, 4
The Solar System
Pergamon, 1984

W J Kaufmann III   1, 8, 9, 10, 11
Relativity and Cosmology
Harper & Row, 1977

W J Kaufmann III   3, 4, 5, 7, 8, 9, 10, 11
Black Holes and Warped Spacetime
Freeman & Co, 1979

C R Kitchin   3, 4, 5, 6, 7, 8
Stars, Nebulae and the Interstellar Medium
Adam Hilger, 1987

R Kippenhahn   2, 9, 10
100 Billion Suns
Weidenfeld and Nicholson, 1983

A J Meadows   3, 4, 5, 6, 7
Stellar Evolution
Pergamon, 1978

S Mitton   2, 9, 10
  Exploring the Galaxies
  Faber and Faber, 1976
P Murdin and L Murdin   7
  Supernovae
  Cambridge University Press, 1985
J V Narlikar   1, 3, 5, 6, 7, 10, 11
  The Structure of the Universe
  Oxford University Press, 1977
J V Narlikar   1, 4, 8, 10
  The Lighter Side of Gravity
  Freeman & Co, 1982
J V Narlikar   1, 2
  The Primaeval Universe
  Oxford University Press, 1988
I K M Nicolson   1, 2, 3, 8, 10
  Gravity, Black Holes and the Universe
  David and Charles, 1981
B Parker   1, 2
  Creation
  Plenum, 1988
M Rowan-Robinson   1
  Cosmology
  Clarendon, 1977
M Rowan-Robinson   1, 8, 9, 10
  Cosmic Landscape
  Oxford University Press, 1979
I L Rozenthal   1, 2, 10, 11
  Big Bang, Big Bounce
  Springer-Verlag, 1983
S L Shapiro and S A Teukolsky   5, 6, 7, 8
  Black Holes, White Dwarfs and Neutron Stars
  Wiley, 1983
J Silk   1, 10, 11
  The Big Bang
  Freeman & Co, 1980
R Smoluchowski, J N Bahcall and M S Matthews   3, 4, 9
  The Galaxy and the Solar System
  University of Arizona Press, 1986
R J Tayler   3, 4, 5, 6, 7
  The Origin of the Chemical Elements
  Wykeham, 1972
R J Tayler   9, 10
  Galaxies: Structure and Evolution
  Wykeham, 1978

K S Thorne, R H Price and D A MacDonald   8
  Black holes, the Membrane Paradigm
  Yale University Press, 1986
J S Trefil   1
  The Moment of Creation
  Charles Scribner, 1983
R V Wagoner and D W Goldsmith   1, 2, 10
  Cosmic Horizons
  Freeman & Co, 1982
S Weinberg   1
  The First Three Minutes
  André Deutsch, 1977
A Wright and H Wright   1, 2, 9, 10, 11
  At the Edge of the Universe
  Ellis Norwood, 1989

Index

Page numbers in **bold** indicate the beginning of a lengthy section on the topic.

Abundance of the elements, 144
Accretion
  Disc, 54, 113, 115, 157, 182
  Stream, 115
Active galactic nuclei *see* AGNs
AGNs, 172
AM Her stars, 105, 106
Angular momentum
  Conservation, 53
  Transfer, 63
Anthropic principle, 23
Anthropocentrism, 164, 171

Background radiation, 10, 11, 17, 40, 186
Baryon-degenerate material, 81, 101
Bipolar outflow, 57, 59
Black dwarf, 188
Black hole, 26, 144, **147**
  Dustbin aspect, 147
  Galactic remnants, 188
  Infall of matter, 181
  Kerr, 151, 156
  Massive, 159, 171
  Mini, 159
  Schwarzschild, 151
    Radius, 151
BL Lac object, 172, 176, 180
Bok globule, 51
Bremsstrahlung radiation, 96

Casimir effect, 15
Chandrasekhar limit, 98, 102, 129, 158
Comet, 107
Compact object, mass limit, 157
Conservation of mass and energy, 8

Core
  Collapse, 129, 134
    Halt, 138
  Contraction, 81
  Iron, formation of, 132
  Rebound, 130, 138
Cosmic string, 36
Cosmology, 2
  Big crunch, 187
  Bounce, 187
  Chaotic space 'foam', 20
  Future, **184**
  Hot big bang, **6**, 19, 184
  Inflation, 186
    Chaotic, **11**
  Steady state theory, 8
Crab nebula, 131
Crooke's radiometer, 62
Cyg-X1, 158

Deuterium, 8
Dimension, 21
  Compactification, 21, 187
Doppler shift, 174, 186

Earth, age of, 44
Ecliptic, 45
Edge problem, 20, 22
Electron, 7
Electron-degenerate material, 81, 101, 114
Energy
  Field, 20
  Nuclear reactions, 65, 73
    CNO cycle, 75
    Proton–proton chain, 74
  Sink, 52
Endothermic reaction, 129

Entropy, 190
Escape velocity, 148
Event horizon, 151
Exothermic reaction, 125

Flatness problem, 16
Flickering, 111
Fluctuation
  Adiabatic, 36
  Isothermal, 36
Forbidden line, 94
Force
  Apparent, 149
  Field, 149
Free–free radiation, 96

Galaxy, **25, 160**
  Andromeda, 26
  Black hole remnant, 188
  Cluster of, 33
  Corona, 168
  Elliptical, 28
  Halo, 168, 169
  Irregular, 28
  Radio, 9, 10, 28, 177, 180
  Recession, 6
  Red-shift, 6
  Seyfert, 28, 29, 172, 176
  Supercluster of, 34
General relativity, 12, 149
  Paradox, 153
Globular cluster, 164
Giant molecular cloud *see* GMC
Giant star, 66
GMC, 46
Gravity
  Lens, 41, 175
  Radiation, 188
  Red-shift, 150
  Time dilation, 151
  Waves, 39, 149
Great attractor, 41
Greenhouse effect, 72, 79
  Runaway, 72, 79

Hawking radiation, 155, 189
Heisenberg uncertainty principle, 13

Helium, 5
  Burning *see* Triple alpha reaction
  Flash, 85
Herbig Ae and Be star, 67
Herbig–Haro object, 57
Hertzsprung–Russell diagram, 65, 83
Horizon distance, 18, 39
Hubble Law, 6
Hydrogen
  Atom, 7
  Fusion reactions, 114
  Heavy, 8
  21 cm radio emission, 166
HI region, 46
HII region, 46

Index notation, 2
Interplanetary medium, 45
Interstellar
  Dust, 48, 61, 165
  Molecules, 48
  Nebula, 45
IRAS spacecraft, 49
Iron, break-up of nucleus, 135
Isotropy problem, **16**

Jeans' mass, 49
Jet, 176

Lamb shift, 15
Large Magellanic Cloud, 119
Lithium excess, 67
Lyman-alpha forest, 175

Magnetic monopoles, 19, 188
Main sequence, 65
  Lifetime, 81
Maser, 49, 69
Meteorite, 70
Milky Way galaxy, **160**
  Centre, 161, 170
    of rotation, 171
  21 cm map, 167
Mira, 98
Missing mass, 16, 30, **185**

Nebula
  Horsehead, 48
  Interstellar, 45
Nebulium, 94
Neutrino, 7, 74, 119
  Absorption, 131
  Opacity, 137
  Pulse, 124, 129, 136
  Solar
    Problem, 76
    Spectrum, 77
  Telescope, 75
  Trapping, 130
Neutron, 7
Nova
  Classical, 106, **107**
  Dwarf, 106, **107**
  Light curve, 112
  Recurrent, **107**
  Shell, 117
  Spectrum, 111
Nucleosynthesis, 65, 85

Olbers' paradox, 4

Penrose process, 157
Planetary nebula, 46, 86, **88**
  Central star, 89, **99**
  Density, 95
  Spectrum, 94, 95
    Radio, 97
  Temperature, 95
Planets, formation of, 63
Polars see AM Her stars
Potential energy, gravitational, 15
Precession, 108
Pressure
  Nucleon, 130
  Radiation, 61
Proton, 7
  Decay, 119, 188
Protostar, **50**
Pulsar, 123, 146
QSO, **172**
  Activity, 183
  Barthel's model, 179

Quantum
  Fluctuation, 14, 20, 187
    Zero-energy, 15
  Mechanics, 12
Quasar, 26, 29, 153, **172**
Quasi-stellar object, see QSO

Ring nebula in Lyra, 89

Sagittarius A*, 171
Saint Augustine, 1
Schönberg–Chandrasekhar limit, 80
Shock wave, 130, 139, 140, 141
Singularity, 12, 151, 152
  Problem, 12
Sirius, 103
Sonic point, 130
Space-time, 149
SS433, 105, 159
Star
  Close binary, 101, 104, 111, 122
    X-ray, 146
  Gauging, 163, 165
  Light, 154
  Neutron, 101, 123
  Pre-supernova, 126
  Rotational velocity, 58, 60
  Spectrum, 34
  Trails, 3
  Variable, 85, 97
  Wind, 99
Static limit, 157
Sun, **72**
  Eclipse, 82
  Lifetime, 78
    Main sequence, 81
  Neutrino problem, 76
  Origin, 44
Superluminal expansion, 177, 178, 181
Supernova, 70
  Light curve, 124
  Optical phase, 143
  Remnant, 46, 145
  Type I, 106, 115, **119**
    Spectrum, 121, 125
  Type II, **119**
    Spectrum, 123, 125

Superstring theory, 21, 36

T Tauri star, 66
Temperature
    Effective, 96
    Kinetic, 96
Tide, 153, 155
Time
    Beginning, 1
    End, 189
Triple alpha reaction, 84, 127

U Gem star, 113
Universe
    Age, 4
    Closed, 186
    Composition, 6
    Density
        Critical, 16, **184**
        Mean, 16
        Parameter (Omega), **185**
    Energy, 15
    Inflation, **18**
    Open, **187**

Vacuum, false, 39
Virtual particle, 14, 155
Void, 35

White dwarf star, 66, 86, 89, 113,
    123
    Structure, 102
Wolf–Rayet star, 99
Wormhole, 155

Z Cam star, 113, 117
Zero-point energy, 13
Zodiacal light, 45